Life Science Grade Five

Table of Contents

Life Science Grade Five

Introduction

We wake up in a new world every day. Our lives are caught in a whirlwind of change in which new wonders are constantly being discovered. Technology is carrying us headlong into the 21st century. How will our children keep pace? We must provide them with the tools necessary to go forth into the future. Those tools can be found in a sound science education. One guidepost to a good foundation in science is the National Science Education Standards. This book adheres to these standards.

Children are naturally curious about science and life. They see the world around them and ask questions that naturally lead into the lessons that they will be taught in science. Science is exciting to children because it answers their questions about themselves and the world around them—their immediate world and their larger environment. A basic understanding of science boosts students' understanding of the world around them.

As children learn more about themselves and their world, they should be encouraged to notice the other living things that inhabit their world. They should be aware of the life cycles of all living things. They should become aware of the interdependence of organisms—from plants, to animals, to humans—and how organisms affect and are affected by their environments. Children should also learn how they can control their own environments to promote their health. Through good personal hygiene, exercise, and good decisions while interacting with nature, children will learn how to take better care of themselves.

Organization

Life Science serves as a handy companion to the regular science curriculum. It is broken into three units: The Cell; Plants and Animals; and Your Body, Inside and Out. Each unit contains concise background information on the unit's topics, as well as exercises and activities to reinforce students' knowledge and understanding of basic principles of science and the world around them.

- **The Cell**: By the fifth grade, students should be aware that the cell is the basic unit of which all living things are made. This unit describes the cell in detail—its needs, appearance, makeup, and purpose. Cell division is discussed, and students look at cells under a microscope. Students are reacquainted with single-celled organisms.
- **Plants and Animals**: Students review the basic needs of plants and animals. They learn about the interdependencies of plants and animals by studying populations, communities, and food chains. Students study the earth's biomes and the importance of preserving them. Students learn how plants and animals can adapt to gradual changes in environment. They also come to understand that adaptations take many years and that organisms cannot adapt to sudden changes in environment. These changes can cause endangerment and extinction of species.
- **Your Body, Inside and Out**: By the fifth grade, students are ready to learn about the human body and its systems. The ten systems of the human body are described, and students gain more understanding through experimentation. Students are encouraged to practice good dietary and hygiene habits and to exercise to keep their bodies healthy. The concept of inherited traits is introduced, and students consider similarities in their own family.

This book contains three types of pages:

- Concise background information is provided for each unit. These pages are intended for the teacher's use or for helpers to read to the class.

- Exercises are included for use as tests or practice for the students. These pages are meant to be reproduced.

- Activity pages list the materials and steps necessary for students to complete a project. Questions for students to answer are also included on these pages as a type of performance assessment. As much as possible, these activities include most of the multiple intelligences so students can use their strengths to achieve a well-balanced learning style. These pages are also meant for reproduction for use by students.

Use

Life Science is designed for independent use by students who have been introduced to the skills and concepts described. This book is meant to supplement the regular science curriculum; it is not meant to replace it. Copies of the activities can be given to individuals, pairs of students, or small groups for completion. They may also be used as a center activity. If students are familiar with the content, the worksheets may also be used as homework.

To begin, determine the implementation that fits your students' needs and your classroom structure. The following plan suggests a format for this implementation.

1. Explain the **purpose** of the worksheets to your students. Let them know that these activities will be fun as well as helpful.

2. Review the **mechanics** of how you want the students to work with the activities. Do you want them to work in groups? Are the activities for homework?

3. Decide how you would like to use the **assessments**. They can be given before and after a unit to determine progress, or only after a unit to assess how well the concepts have been learned.

4. Determine whether you will send the tests home or keep them in students' **portfolios**.

5. Introduce students to the **process** and the purpose of the activities. Go over the directions. Work with students when they have difficulty. Work only a few pages at a time to avoid pressure.

6. Do a **practice** activity together.

The Scientific Method

Students can be more productive if they have a simple procedure to use in their science work. The scientific method is such a procedure. It is detailed here, and a reproducible page for students is included on page 7.

1. **PROBLEM**: Identify a problem or question to investigate.
2. **HYPOTHESIS**: Tell what you think will be the result of your investigation or activity.
3. **EXPERIMENTATION**: Perform the investigation or activity.
4. **OBSERVATION**: Make observations, and take notes about what you observe.
5. **CONCLUSION**: Draw conclusions from what you have observed.
6. **COMPARISON**: Does your conclusion agree with your hypothesis? If so, you have shown that your hypothesis was correct. If not, you need to change your hypothesis.
7. **PRESENTATION**: Prepare a presentation or report to share your findings.
8. **RESOURCES**: Include a list of resources used. Students need to give credit to people or books they used to help them with their work.

Hands-On Experience

An understanding of science is best promoted by hands-on experience. *Life Science* provides a wide variety of activities for students. But students also need real-life exposure to their world. Playgrounds, parks, zoos, aquariums, and even vacant lots are handy study sites to observe many organisms. Repeated visits to the same site can help to show students that the organisms are constantly changing.

It is essential that students be given sufficient concrete examples of scientific concepts. Appropriate manipulatives can be bought or made from common everyday objects. Most of the activity pages can be completed with materials easily accessible to the students.

Science Fair

Knowledge without application is wasted effort. Students should be encouraged to participate in their school science fair. To help facilitate this, each unit in *Life Science* ends with a page of science fair ideas and projects. Also, on page 8 is a chart that will help students to organize their science fair work.

To help students develop a viable project, you might consider these guidelines:

- Decide whether to do individual or group projects.

- Help students choose a topic that interests them and that is manageable. Make sure a project is appropriate for a student's grade level and ability. Otherwise, that student might become frustrated. This does not mean that you should discourage a student's scientific curiosity. However, some projects are just not appropriate. Be sure, too, that you are familiar with the school's science fair guidelines. Some schools, for example, do not allow glass or any electrical or flammable projects. An exhibit also is usually restricted to three or four feet of table space.

- Encourage students to develop questions and to talk about their questions in class.

- Help students to decide on one question or problem.

- Help students to design a logical process for developing the project. Stress that the acquisition of materials is an important part of the project. Some projects also require strict schedules, so students must be willing and able to carry through with the process.

- Remind students that the scientific method will help them to organize their thoughts and activities. Students should keep track of their resources used, whether they are people or print materials. Encourage students to use the Internet to do research on their project.

Additional Notes

- **Parent Communication**: Send the Letter to Parents home with students so that parents will know what to expect and how they can best help their child.

- **Bulletin Board**: Display completed work to show student progress.

- **Portfolios**: You may want your students to maintain a portfolio of their completed exercises and activities, or of newspaper articles about current events in science. This portfolio can help you in performance assessment.

- **Assessments**: There are assessments for each unit at the beginning of the book. You can use the tests as diagnostic tools by administering them before children begin the activities. After children have completed each unit, let them retake the unit test to see the progress they have made.

- **Center Activities**: Use the worksheets as a center activity to give students the opportunity to work cooperatively.

- **Have fun**: Working with these activities can be fun as well as meaningful for you and your students.

Curriculum Correlation

Curriculum Area	Page Numbers
Social Studies	136, 137
Language Arts	14, 15, 17, 22, 23, 27, 30, 31, 39, 40, 41, 44, 46, 48, 49, 51, 52, 53, 55, 61, 62, 63, 69, 71, 72, 73, 74, 76, 77, 78, 83, 84, 87, 88, 89, 90, 91, 92, 93, 94, 95, 107, 110, 113, 118, 120, 121, 126, 128, 130, 141
Math	19, 45, 58, 62, 63, 68, 85, 86, 92, 96, 114, 117, 129, 131, 132
Physical Education/Health	123, 124, 125, 139, 140, 141
Art	20, 21, 24, 25, 26, 28, 32, 41, 52, 54, 56, 75, 79, 122, 127

FOSS Correlation

The Full Option Science System™ (FOSS) was developed at the University of California at Berkeley. It is a coordinated science curriculum organized into four categories: Life Science; Physical Science; Earth Science; and Scientific Reasoning and Technology. Under each category are various modules that span two grade levels. The modules for this grade level are highlighted in the chart below.

Module	Page Numbers
Food and Nutrition	12-13, 47, 48, 94-97, 99, 100, 101, 104, 105-106, 126, 127, 133, 134, 139-140, 141
Environments	34-38, 39, 43, 44, 46, 48, 49-50, 65, 74, 75, 76, 78-79, 80, 81-82, 83-84, 85-86, 87-88, 89-90, 91-92

Dear Parent,

During this school year, our class will be using an activity book to reinforce the science skills that we are learning. By working together, we can be sure that your child not only masters these science skills but also becomes confident in his or her abilities.

From time to time, I may send home activity sheets. To help your child, please consider the following suggestions:

- Provide a quiet place to work.
- Go over the directions together.
- Help your child to obtain any materials that might be needed.
- Encourage your child to do his or her best.
- Check the activity when it is complete.
- Discuss the basic science ideas associated with the activity.

Help your child to maintain a positive attitude about the activities. Let your child know that each lesson provides an opportunity to have fun and to learn more about the world around us. Above all, enjoy this time you spend with your child. As your child's science skills develop, he or she will appreciate your support.

Thank you for your help.

Cordially,

Name ______________________________ Date ______________

THE SCIENTIFIC METHOD

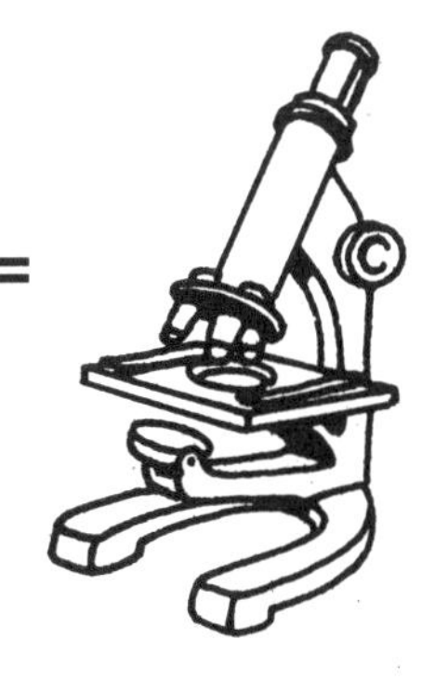

Did you know you think and act like a scientist? You can prove it by following these steps when you have a problem. These steps are called the scientific method.

1. **PROBLEM**: Identify a problem or question to investigate.

   ____

   ____

2. **HYPOTHESIS**: Tell what you think will be the result of your investigation or activity.

   ____

   ____

3. **EXPERIMENTATION**: Perform the investigation or activity.

   ____

4. **OBSERVATION**: Make observations, and take notes about what you observe.

   ____

   ____

   ____

5. **CONCLUSION**: Draw conclusions from what you have observed.

   ____

   ____

   ____

6. **COMPARISON**: Does your conclusion agree with your hypothesis? If so, you have shown that your hypothesis was correct. If not, you need to change your hypothesis.

   ____

7. **PRESENTATION**: Prepare a presentation or report to share your findings.
8. **RESOURCES**: Include a list of resources used. You need to give credit to people or books you used to help you with your work.

   ____

   ____

   ____

Name ______________________ Date ______________

THE SCIENCE FAIR

The science fair at your school is a good place to show your science skills and knowledge. Science fair projects can be several different types. You can do a demonstration, make a model, present a collection, or perform an experiment. You need to think about your project carefully so that it will show your best work. Use the scientific method to help you to organize your project. Here are some other things to consider:

Project Title ______________________

Working Plan	**Date Due**	**Date Completed**	**Teacher Initials**
1. Select topic			
2. Explore resources			
3. Start notebook			
4. Form hypothesis			
5. Find materials			
6. Investigate			
7. Prepare results			
8. Prepare summary			
9. Plan your display			
10. Construct your display			
11. Complete notebook			
12. Prepare for judging			

Write a brief paragraph describing the hypothesis, materials, and procedures you will include in your exhibit.

Be sure to plan your project carefully. Get all the materials and resources you need beforehand. Also, a good presentation should have plenty of visual aids, so use pictures, graphs, charts, and other things to make your project easier to understand.

Be sure to follow all the rules for your school science fair. Also, be prepared for the judging part. The judges will look for a neat, creative, well-organized display. They will want to see a clear and thorough presentation of your data and resources. Finally, they will want to see that you understand your project and can tell them about it clearly and thoroughly. Good luck!

Name ______________________________ Date ______________

UNIT 1 ASSESSMENT: THE CELL

Directions: Match the terms at the right with the definitions at the left.

_____ **1.** one of the major theories of life science

_____ **2.** the movement of a material into an area that has less of the material

_____ **3.** loss of water

_____ **4.** organelles that make food in plant cells

_____ **5.** the living material between a cell's nucleus and its cell membrane

_____ **6.** the basic unit of structure and function in an organism

_____ **7.** structures in the nucleus of a cell that contain the instructions that enable the nucleus to control the activities of the cell

_____ **8.** organelles that produce the energy a cell needs

_____ **9.** a protective covering around a cell

_____ **10.** the control center of a cell

_____ **11.** any cell structure that has a specific job to do

_____ **12.** the diffusion of water through a cell membrane

a. cell

b. cell membrane

c. cell theory

d. chloroplasts

e. chromosomes

f. cytoplasm

g. dehydration

h. diffusion

i. mitochondria

j. organelle

k. osmosis

l. nucleus

Name ______________________________ Date ________________

Unit 2 Assessment: Plants and Animals

Directions: Use the words from the box to complete the sentences.

imprints	cold-blooded	vertebrates	arthropods
warm-blooded	amphibians	bones	carbon dioxide
mammals	photosynthesis	angiosperms	invertebrates
metamorphosis	food chain		

1. Plants take in ____________________.
2. Animals that live part of their lives in water and part on land are called ____________________.
3. Animals with backbones are called ____________________.
4. ____________________ is the process by which plants make food.
5. Animals with hair or fur are called ____________________.
6. Plants that reproduce by seeds in cones are ____________________.
7. A frog's development is called ____________________.
8. Most birds have hollow ____________________.
9. A ____________________ shows how organisms depend upon each other for food.
10. Animals whose body temperature changes with that of the surrounding water or air are ____________________.
11. Animals without backbones are called ____________________.
12. Animals whose body temperature remains steady are ____________________.
13. Fossils that are impressions left in sand or mud are called ____________________.
14. Insects are the largest part of a group of animals called ____________________.

Name ______________________________ Date ______________

Unit 3 Assessment: Your Body, Inside and Out

Directions: Choose the correct words from the box to complete the paragraphs. Write the number in the space next to each word to indicate where it belongs.

_____ tissue	_____ germs	_____ system	_____ hygiene
_____ involuntary	_____ exercise	_____ toned	_____ blood
_____ joints	_____ circulatory	_____ flossing	_____ Red
_____ cell	_____ White	_____ voluntary	_____ organ
_____ muscles	_____ brushing	_____ lungs	_____ systems
_____ brain	_____ bones	_____ respiratory	
_____ senses	_____ nervous	_____ carbon dioxide	

The (1) is the basic building block of all living things. Groups of cells that have the same structure and do the same job are called (2). An (3) is a group of tissues working together to do a specific job. A (4) is a group of organs working together to do a job. There are ten (5) in the human body. The (6) system controls all the others.

We have five (7) that we use to collect information from the world around us. The information that we collect is sent to the (8), which in turn sends messages to the body telling it what to do.

Our (9) give us support, our (10) make our bodies move, and (11) allow our bones to bend. We need to (12) our body in order to keep our muscles firm, or (13). Exercise sends (14) to the muscle cells. Some muscles are (15) muscles. We do not think about moving them, but they move anyway. Others are (16) muscles. We can move them when we want to.

The (17) system and the (18) system work together to bring oxygen and nutrients to the cells of the body. Oxygen is collected in the (19). The cells get rid of (20) there, too. (21) blood cells carry oxygen and nutrients to the other cells in the body. (22) blood cells fight infection.

Along with exercise, we need to practice good (23) to stay healthy. This means (24) and (25) our teeth and keeping our body clean. We also need to prevent the spread of (26) from one person to another. We need to take good care of our body so that it will last a long time!

Unit 1: The Cell
Background

Cells

In the 1600s, Robert Hooke identified and named the cell. All living things are made up of cells, the smallest living units. The cell has all of the properties of a living thing. A cell grows, reproduces, consumes energy, changes it, and excretes waste. Cells react to stimuli and to changes in the environment. Most cells contain a nucleus, cytoplasm, and a cell membrane. Plant cells have, in addition, a cell wall, which is outside of the membrane. The cell wall is made of cellulose and makes the plant cell stiffer than the animal cell. A tree trunk is hard because of the cell walls in the plant cells. The crunch of a carrot or celery is caused by the cell walls breaking as we bite into the food.

The nucleus is the control center of the cell, and it is where the cell begins reproduction. The nucleus contains the chromosomes that determine which hereditary traits are passed on from parents to offspring. The chromosomes are made up primarily of DNA. DNA molecules can duplicate themselves. The cytoplasm is where the special functions of the cell are carried on. The structure and role of the cytoplasm change from one type of cell to another. However, the nucleus (with its DNA) controls what the special function of the cytoplasm will be. The nucleus gives the cytoplasm what it needs to perform its special function. Within the cytoplasm are other structures called *organelles*. An organelle is any part of a cell with a specific job to do. One of these organelles is the mitochondria. The mitochondria supply the cell with the energy it needs. Other organelles are the ribosomes, the endoplasmic reticulum, and the vacuoles. Ribosomes provide protein. The endoplasmic reticulum carries materials to and from the membrane, and the vacuoles carry food and water throughout the cell. The cell membrane gives the cell its shape and controls what passes into and out of the cell. There are many different types of cells, and each performs specific functions. The human body alone has many different types of cells inside it. These types of cells will be explored in Unit 3.

Besides the cell wall, many plant cells have another unique component—chloroplasts. The chloroplasts are small, green bodies shaped like footballs that can be found in the cytoplasm of a plant cell. The chloroplasts give the plant its green color and are vital to the process of photosynthesis. Without the chloroplasts, a plant would not be able to make its own food. Photosynthesis takes place inside the chloroplasts. The parts of a plant that are not green, for example, the bulb of an onion, do not have chloroplasts.

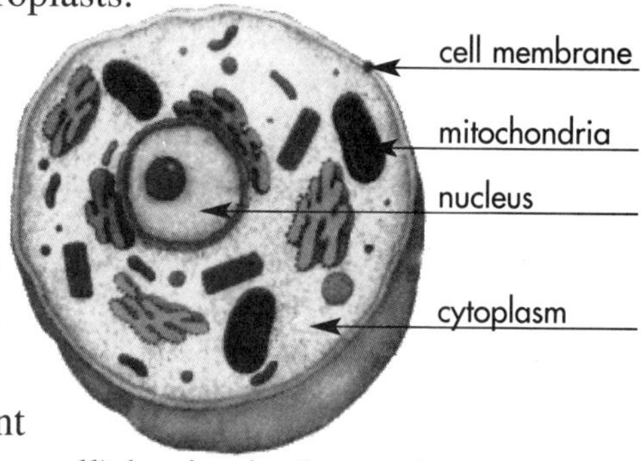

Cell Division

The division of an animal cell differs somewhat from the division of a plant cell. Division of a plant cell (or a cambium cell) begins in the nucleus, as does the division of animal cells. The nucleus becomes two nuclei. But then a new cell wall grows inside the plant cell that divides the cell in two. Then each new cell grows until it is about the same size as the old cell.

The division of a cell is called *mitosis*. Before a cell divides, it enters a stage called *interphase* during which it digests food, uses it for energy, and excretes waste. During this time the cell grows in preparation for division. Different cells divide at different rates, from 15 to 30 minutes to almost two days. The growth of a living thing is caused by the growth and division of its cells.

UNIT 1: THE CELL
BACKGROUND (CONTINUED)

There are four stages of cell division.

- The first, *prophase*, is when the chromosomes in the nucleus become shorter and fatter, then duplicate themselves. The membrane of the nucleus begins to break down.
- During the second phase, *metaphase*, the chromosomes line up across the nucleus of the cell. Each chromosome begins to pull apart, separating the duplicated information from the original information.
- At *anaphase*, the third phase, the chromosomes pull toward opposite ends of the cell.
- During the last phase, *telophase*, cytoplasm is divided, and the nucleus reorganizes into two cells.

Cell Needs

Cells need water, nutrients, and other materials to function. Some of what the cell needs passes in and out of the cell by diffusion, the movement of materials from an area with a lot of the material to an area with less of the material. Water moves across the cell membrane by osmosis. If a cell does not get enough water, or loses more than it takes in, dehydration will occur. When a person sweats, or a plant wilts, water needs to be added to the dehydrated cells.

Single-Cell Organisms

Scientists have classified life into three categories—plant, animal, and protist. The protists do not easily fall within either the plant or the animal category. These tiny one-celled organisms exhibit some of the characteristics of each. The euglena, for example, has chloroplasts in which photosynthesis takes place in the presence of light. However, if there is no light, the euglena will look for food like an animal. The amoeba is another protist. It changes its shape constantly, as it moves one part of its body forward and the rest of its body follows. It also surrounds its food this way. The paramecium and the stentor both have cilia, tiny hair-like projections that are used for swimming and for propelling food into the organism's mouth. The paramecium is identified by its slipper-like shape. The stentor is shaped like a trumpet.

These single-celled organisms can be seen in pond water. After a jar of pond water is collected, students may add a few grains of rice or some hard-boiled egg. After the jar has sat for a few days, the water will be cloudy and full of life. Bacteria will have used the food to grow and reproduce, and other single-celled organisms will consume the bacteria. All of these organisms will be moving about in the water, eating and reproducing.

Name ______________________ Date ______________

CELLS

Read the paragraph, then answer the questions that follow.

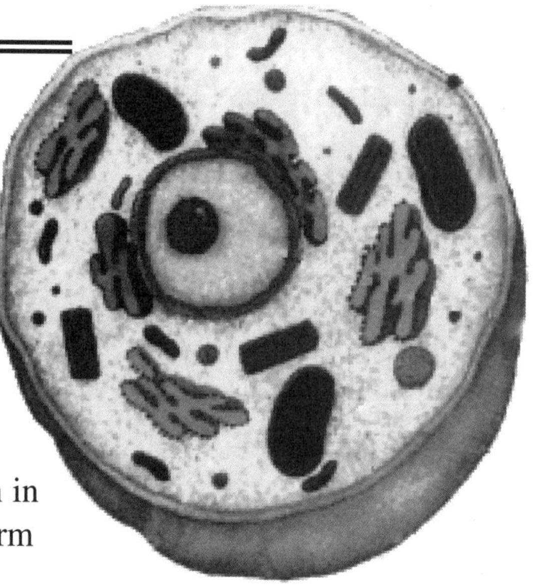

Robert Hooke, working with a microscope and some cork in the 1600s, was the first scientist to identify and name the basic unit of living things—the cell. Hooke's discovery led to the development of the cell theory, with its three parts. First, all living things are made up of cells. Second, the cell is the smallest unit of structure and function in all living things. Third, all cells can reproduce to form new cells.

1. Who was the first person to use the term *cell*? ______________________

2. In what material did he first see cells? ______________________

3. What are the three parts of the cell theory?

a. ______________________

b. ______________________

c. ______________________

Name ______________________ Date ______________

WHAT MAKES UP A CELL?

Read the paragraph, then complete the diagram that follows.

Although cells are the basic units of living things, they are made up of many parts, each with a specific function. All cells are surrounded by a cell membrane that controls what goes into and out of the cell. The nucleus is the command center of a cell. It controls everything that goes on inside the cell. Between the cell membrane and the nucleus is a thick liquid called the cytoplasm. Suspended within the cytoplasm are other structures—organelles. An organelle is any cell structure with a specific job to do. For example, mitochondria supply a cell with the energy it needs.

Label the cell parts on the diagram below.

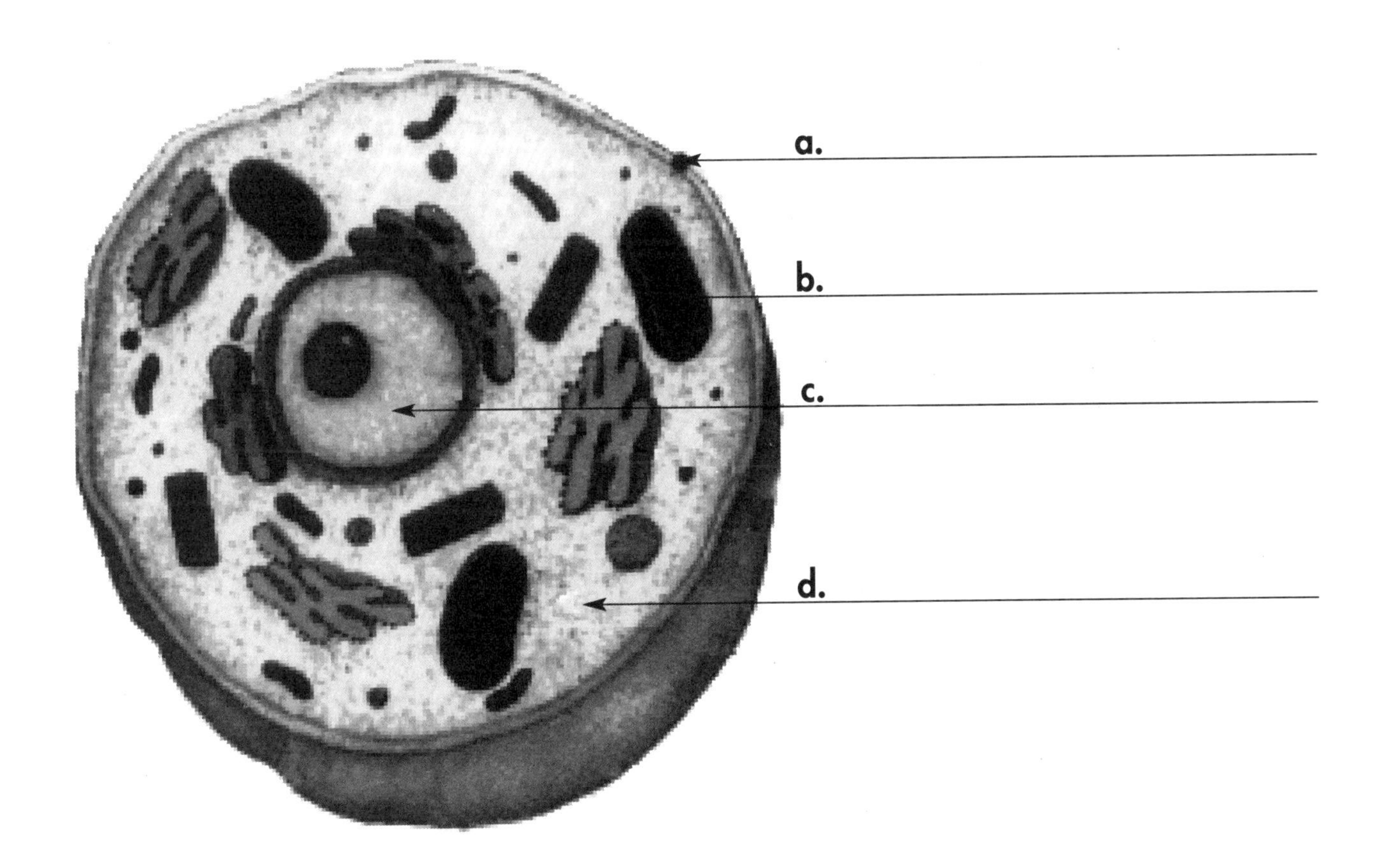

Name ______________________________ Date ______________

PLANT AND ANIMAL CELLS

Plant cells are different from animal cells. Plant cells have a cell wall that makes the plant cell stiffer than an animal cell. Plant cells also have small, green bodies shaped like footballs in the cytoplasm. These are chloroplasts. Chloroplasts are responsible for photosynthesis. Without chloroplasts, plant cells could not make their own food.

A. Look at the pictures. Write the number or numbers next to the term that matches each cell part.

______ cell membrane ______ cytoplasm ______ cell wall

______ chloroplasts ______ nucleus

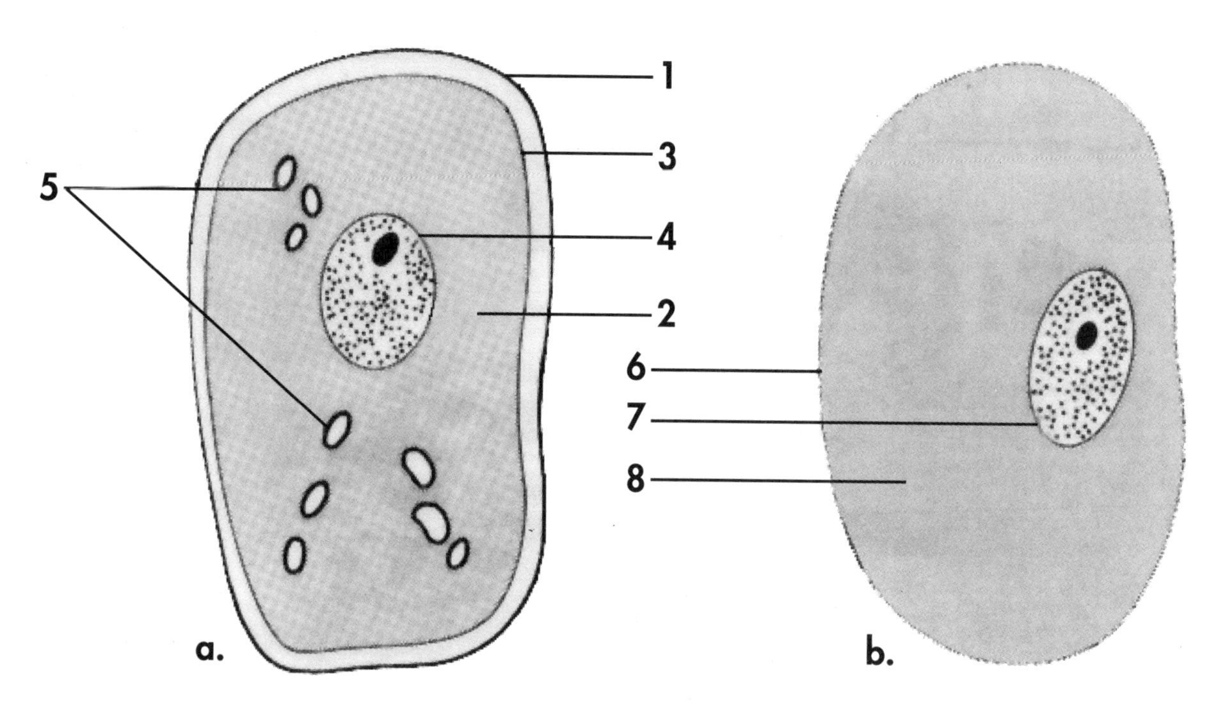

B. Answer these questions.

1. Which two parts are found only in plant cells? ______________________
2. In what cell parts does photosynthesis take place? ______________________
3. What part of a plant keeps the plant stiff? ______________________
4. What cell part allows only certain substances to diffuse into and out of the cell? ______________________
5. Which cell is a plant cell? ______________________

Name ______________________________ Date ______________

WHAT DO CELLS NEED?

Read the paragraph, then answer the questions that follow.

Cells need water, nutrients, and other materials to function. Some materials move into and out of cells by diffusion, the movement of materials from an area with a lot of the material to an area with less of the material. Water moves across cell membranes by a kind of diffusion called osmosis. If a cell loses more water than it takes in, dehydration occurs.

1. Fill in the blanks with the following terms that describe the movement of materials into and out of cells: *cell membrane*, *diffusion*, *nucleus*, *osmosis*, *dehydration*.

______________________ is the movement of materials from an area that has a lot of the material to an area that has less of the material.

______________________ happens when cells lose more water than they take in.

______________________ is the diffusion of water through a cell membrane.

The ______________________ controls the movement of materials, since it controls all the activities of a cell.

The ______________________ selects what passes into and out of a cell.

2. Explain what you think happens to the cells of your body if you sweat a lot but do not drink enough water.

____

____

____

3. Explain what happens when you water a wilted plant.

____

____

____

____

Name ______________________ Date ______________

WORKING CELLS

Cell organelles are any part of a cell with a specific function.

Using a science book or encyclopedia, answer questions 1-4.

1. What organelle surrounds and protects the nucleus? ______________________

2. What organelles act as protein factories for the cells? ______________________

3. What organelle carries materials to and from the membrane? ______________________

4. What organelles carry food and water throughout a cell? ______________________

Within the nucleus of a cell are thin strands called chromosomes, which contain the instructions for controlling all functions of the cell. Before a cell divides, it makes an exact copy of the chromosomes. During cell division, the chromosomes and the cytoplasm of the cell divide so that each new cell receives a full set of chromosomes and half the cytoplasm of the original cell.

5. Look up cell division in a science book or encyclopedia. In the boxes below, draw a cell with two chromosomes, and show what happens to the chromosomes during cell division.

Name ______________________ Date ______________

GRAPH A CELL

Below is a grid and coordinates. The coordinates are given so that you can draw a cell. Do this by placing a dot on the empty grid at the position indicated by the first set of coordinates. Then make a dot for the second set of coordinates. Then go on to the third set. Connect the second dot to the third dot. Continue until all dots are connected. Do the cell membrane first, and then go on to the other structures.

Cell Membrane

1. (1, H) **2.** (1, F) **3.** (6, A) **4.** (10, A) **5.** (15, F) **6.** (15, H) **7.** (10, M) **8.** (6, M) **9.** (1, H)

Nucleus

1. (7, I) **2.** (6, H) **3.** (6, G) **4.** (7, F) **5.** (8, F) **6.** (9, G) **7.** (9, H) **8.** (8, I) **9.** (7, I)

Mitochondria

1. (6, K) **2.** (4, I) **3.** (4, H) **4.** (6, J) **5.** (6, K)

Organelle

1. (10, F) **2.** (8, D) **3.** (9, C) **4.** (11, E) **5.** (11, F) **6.** (10, F)

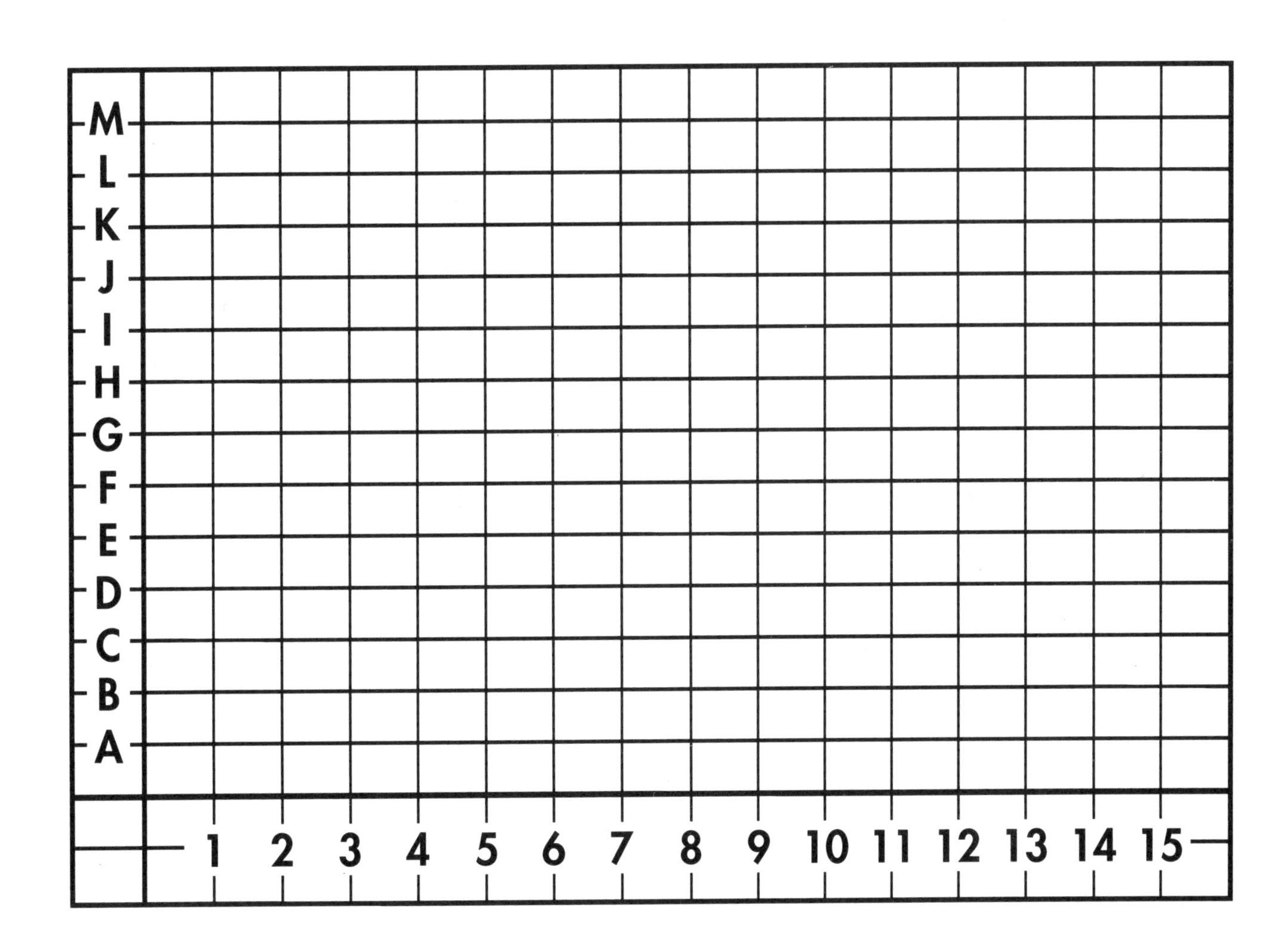

Name ______________________________ Date ______________

LOOKING AT CORK

Robert Hooke learned about cells by looking at a thin slice of cork under a microscope. See what you can learn by looking at the same thing! Look for cells in cork and in other materials you can find.

Materials:
- very thin slice of cork
- microscope slide
- dropper bottle of water
- microscope
- leaves

Do This:
1. Place a very thin slice of cork in a drop of water on a microscope slide.
2. Place the slide under the microscope and adjust the focus.
3. In the space below, draw what you see. Label the cells that Robert Hooke viewed.

4. Break off a small piece of a leaf and put it in a drop of water on a microscope slide.
5. Place the slide under the microscope and adjust the focus.

Go on to the next page.

Name ______________________________ Date ______________

LOOKING AT CORK, P. 2

6. In the space below, draw and label what you see.

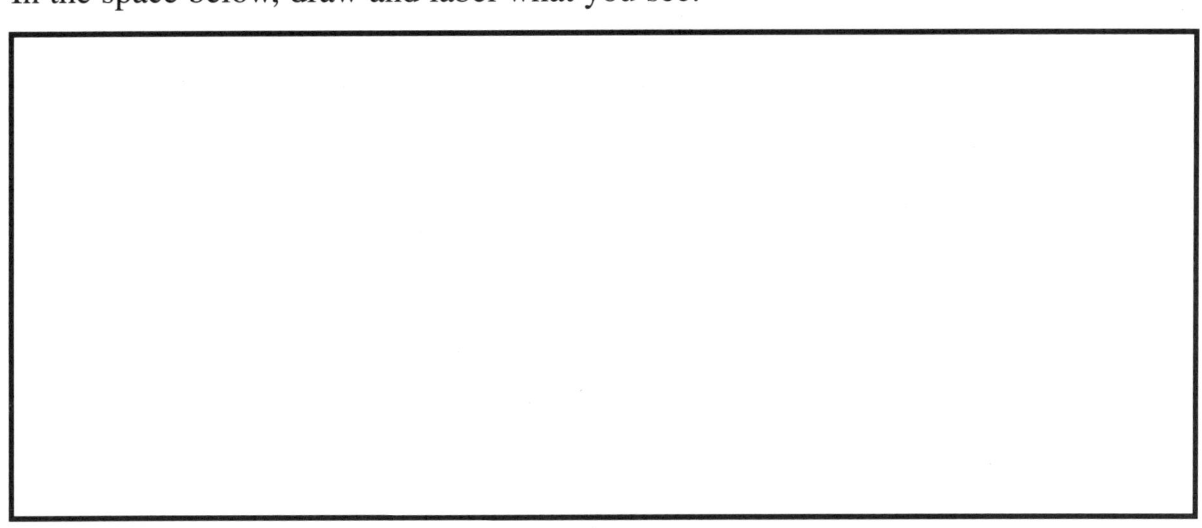

7. Try looking at other things under the microscope. Remember always to use a very thin piece of whatever you are looking at. In the space below, draw and label what you see. Can you see cells? If you can, label the parts.

8. Did you find cells in everything you observed? Explain.

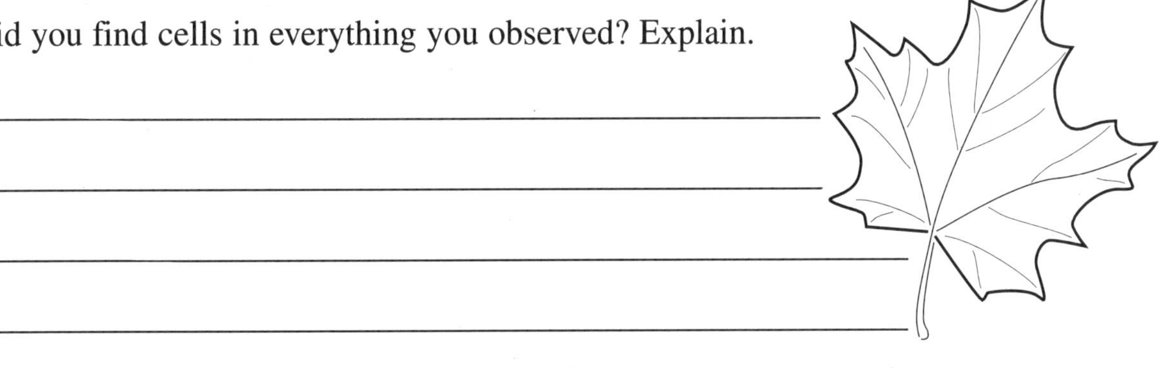

Name ______________________________ Date ______________

COMPARING PLANT AND ANIMAL CELLS

When you compare objects or events, you look for what they have in common. You also look for differences between them.

Decide which cell structures listed below are found in plant cells and which structures are found in animal cells.

nucleus mitochondria chloroplasts cytoplasm organelles cell wall

In the space below, there are two circles that overlap each other. Look at the list of cell structures again. If a structure can be found only in animal cells, write the name of the structure in the part of the "animal cell" circle that does not intersect with the other circle. If a structure can be found in both plant and animal cells, write the name of the structure in the overlapping part of the circles. If a structure can be found only in plant cells, write the name of that structure in the part of the "plant cell" circle that does not overlap the other circle.

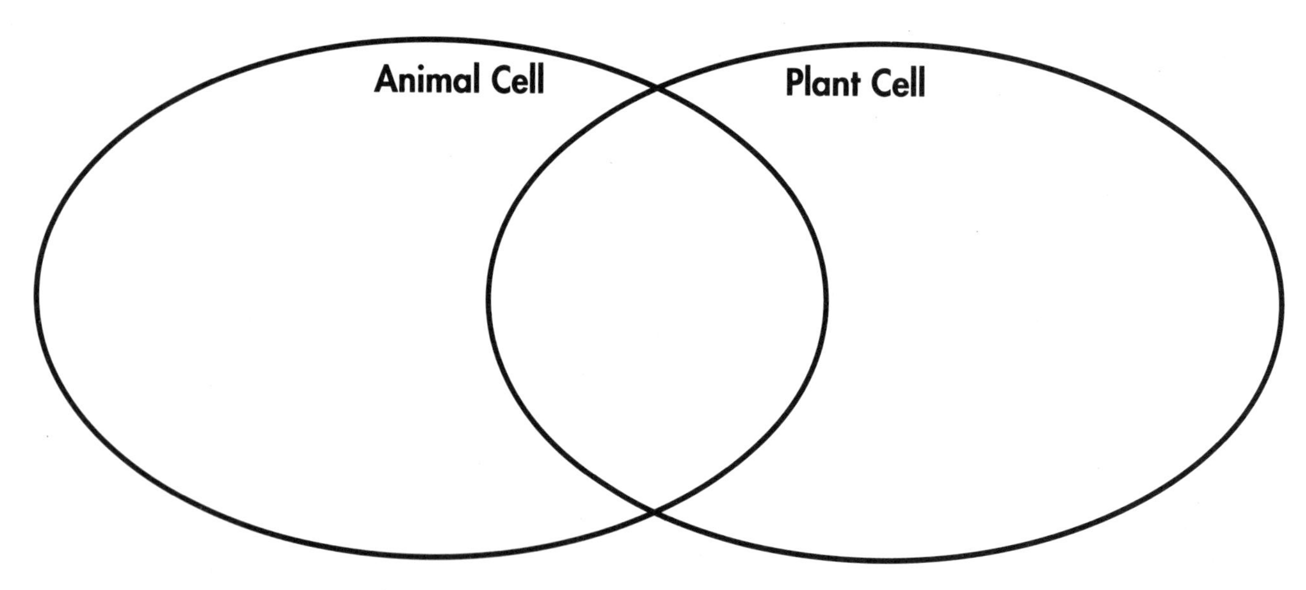

Go on to the next page.

Name ______________________ Date ____________

Comparing Plant and Animal Cells, p. 2

1. What does your diagram tell about the structures in plant and animal cells?

____

____

____

2. Why is using the diagram a good way to help you compare this type of information?

____

____

____

3. Why is it useful to know how plant cells and animal cells are similar and how they are different?

____

____

____

4. How does knowing that plant cells have chloroplasts and that animal cells do not help you compare the way plants and animals get food?

____

____

____

5. How did you use your knowledge of comparisons to help you make good decisions about comparing plant and animal cells?

____

____

____

Name ______________________________ Date ______________

Cells, the Building Blocks of Life

A carrot is very different from an elephant. You seldom see orange elephants, and carrots never eat peanuts. It's easy to tell the difference between these two living things just by looking at them. They have very different characteristics. But one important difference between them can be seen only when you study their cells under a microscope. Carrot cells are very different from elephant cells.

Make a labeled diagram that shows how a cell from a carrot and a cell from an elephant would be different. Then fill in the missing words below each diagram.

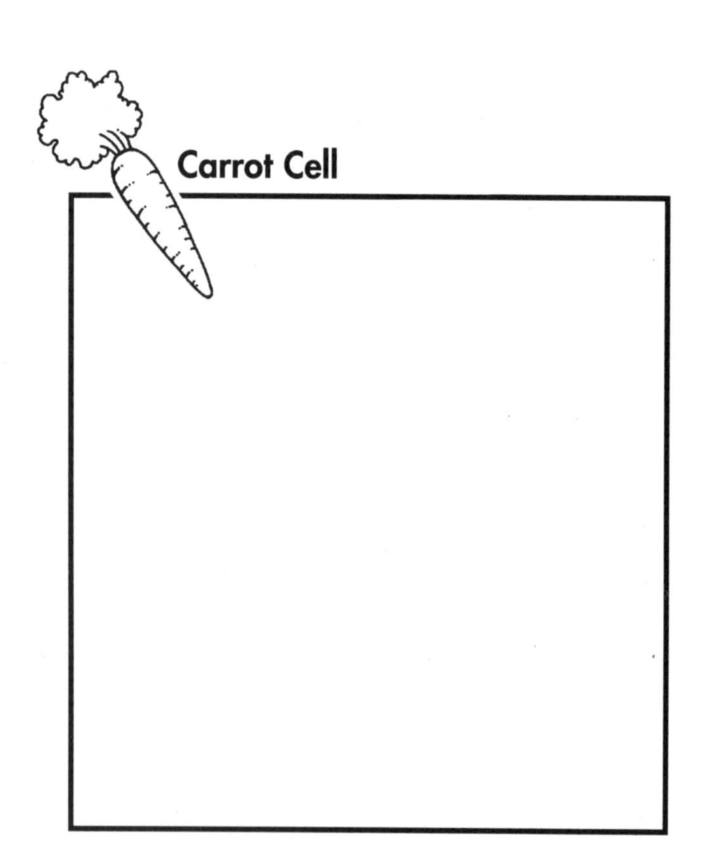

A carrot cell is a

______________________ cell. It has green material called

______________________ in it that helps the carrot make food. The layer outside its cell membrane is its

______________________.

An elephant cell is an

______________________ cell. It has a jellylike liquid inside it called

______________________. Food and water pass into this through its

______________________. The control center of the cell is its

______________________.

Name ______________________________ Date ______________

What Can You Observe In Onion Skin Cells?

You have studied the parts of a plant cell. In this activity, you will be able to see these parts in a thin piece of onion skin containing cells.

Materials:

- piece of onion
- a microscope slide
- forceps (tweezers)
- a cover slip
- eyedropper
- iodine
- microscope

Do This:

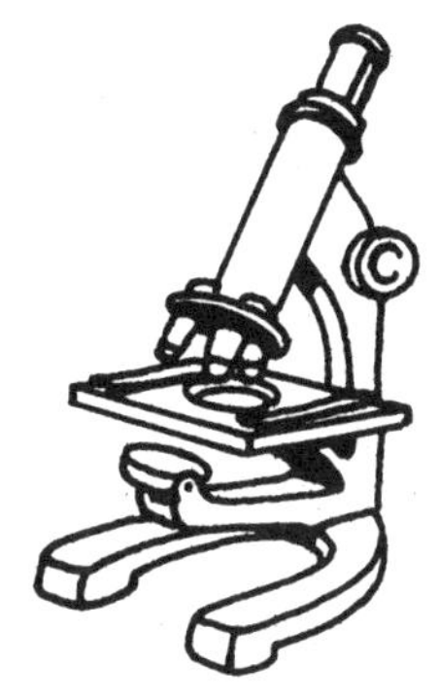

1. Soak the onion slice in water for a few minutes. Use the forceps to gently separate the layers of the onion. Pull apart a small section that is about the size of your fingernail.
2. Place a drop of water on the slide, and put the piece of onion skin on it. Place the cover slip on the onion skin.
3. Place the slide on the microscope. Move the barrel of the microscope up until you get the cells in focus. You may also have to move the slide around to find the thinnest part of the onion skin.

Make a drawing of what you see.

4. Some structures of cells are hard to see even using a microscope. To make them visible, scientists sometimes add something that stains the cells. Then the structure becomes visible. You can stain your onion cells by putting a drop of iodine on the onion skin slide.

Describe what you observed after you stained the onion skin. ______________________________

____

____

____

Name ______________________ Date ____________

What Do Some Kinds of Protists Look Like?

Protists are tiny one-celled organisms that do not fit into either the animal or plant category. They show some characteristics of both plants and animals. This activity will show you what some protists look like.

Materials:

- **jar of pond water**
- **cover slips**
- **eyedropper**
- **microscope**
- **microscope slides**

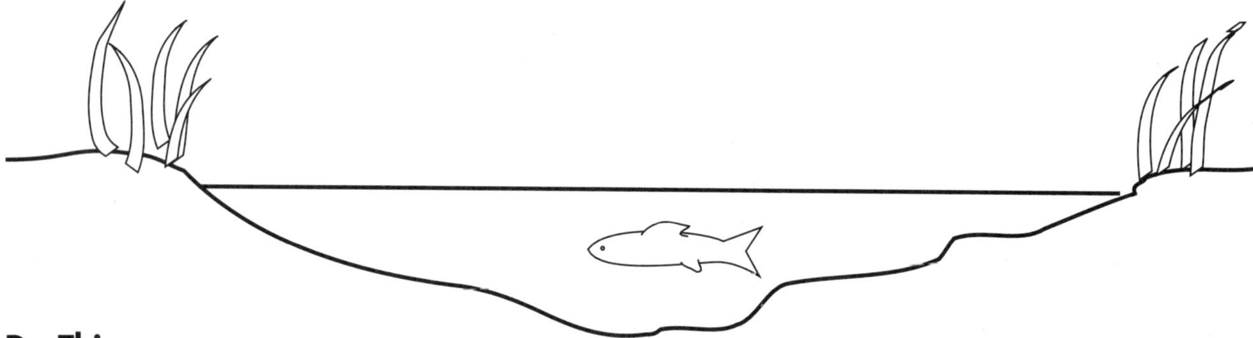

Do This:

1. Put a drop of pond water on a microscope slide. Place a cover slip over the drop of water.
2. Put the slide on the microscope stage.
3. Move the objective (bottom lens) down until it nearly touches the cover slip. Then, look through the microscope.
4. Raise the objective slowly until you can see clearly into the drop of water.
   - **Caution: Do not lower the objective while you are looking through the microscope; you might break the slide or lens.**
5. Look carefully for protists swimming in the drop of water. Gently move the slide around as you look. Be careful not to lift the slide.
6. Draw the protists that you see. Label any parts of cells that you can identify.

Go on to the next page.

Name ______________________ Date ____________

What Do Some Kinds of Protists Look Like?, p. 2

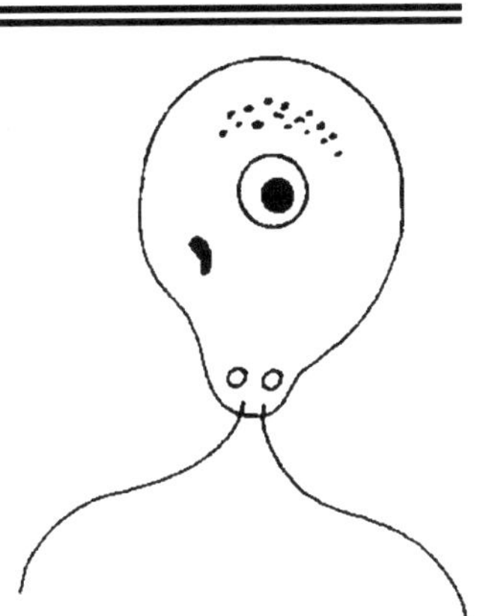

1. How many different kinds of protists did you see? Can you identify any of them?

____

____

____

____

____

2. How would you describe the shape of the protists you saw?

____

____

____

____

3. How did the protists move?

____

____

____

____

4. What conclusions can you make about the colors, sizes, shapes, and structures of protists?

____

____

____

____

____

Name ______________________ Date ____________

THE SIMPLEST ORGANISMS

How good a detective are you? Here is a game called "What Am I?"

Read the clues and then make a drawing of the organism described. Write the name of the organism beside each drawing. Use a science book or an encyclopedia to help you.

1. I am only one cell. I am an animallike protist. I have no regular shape. What am I?

   ____

2. I am only one cell. I am an animallike protist. I have a definite shape. I am shaped like a slipper. What am I?

   ____

3. I am only one cell. I have both animallike and plantlike characteristics. I have a definite shape. What am I?

   ____

4. I am shaped like a crystal. I show signs of life only when I am inside other living cells. What am I?

   ____

Name ______________________ Date ______________

WATER-DROP MICROSCOPE

Scientists use microscopes to study small objects like cells. A microscope magnifies an object so that it appears larger. You can make a small microscope to use at home or just about anywhere.

Materials:

- water
- cotton swab
- aluminum pie pan
- newspaper or magazine
- small nail

Do This:

1. Carefully push the nail slowly and firmly into the bottom of the pie pan until it makes a small, round hole. Put a thick newspaper or magazine under the pie pan to protect the table top.

2. Wet the cotton swab with water. Carefully let a drop of water fall from the cotton swab into the hole. The drop should remain in the hole, not drip down through it. If you have made the hole too big, make a smaller hole in another spot in your pie pan and try again.

3. Choose an object to look at through your microscope. A piece of newspaper with small print, a patterned scarf or handkerchief, or a coin should work well. Hold the pan just above the object. Look at the object through the water drop. Look at several other objects. How do they look? ______________________

You have made a water-drop microscope. The drop of water acts like one of the lenses of an actual microscope.

Name ________________________________ Date ________________

More One-Celled Organisms

Read this story about one-celled organisms that live in ponds. Then follow the directions.

Chlamydomonas are one-celled green plants. They are egg-shaped and have two whiplike threads. When these threads flip back and forth, the chlamydomonas move through the water. During the summer, these organisms reproduce very quickly. They can make a whole pond look bright green.

Closterium are long, narrow, and curved. They look like macaroni that are pointed at the ends. Although closterium have chloroplasts and make their own foods, they are yellow-brown in color. Iron in their cell walls gives them this color.

Didinium are round but pointed at one end. They have two bands of cilia. One band is near the pointed end. The other is around the middle. Didinium are smaller than paramecia. However, this doesn't stop them from eating up to six paramecia a day.

Diatoms have hard glasslike shells. The shells are in two parts, like a box and its lid. There are over 5,000 different kinds of diatoms. Diatoms come in a variety of shapes—triangles, circles, stars, and more. Diatoms are gold and brown. They make their own food.

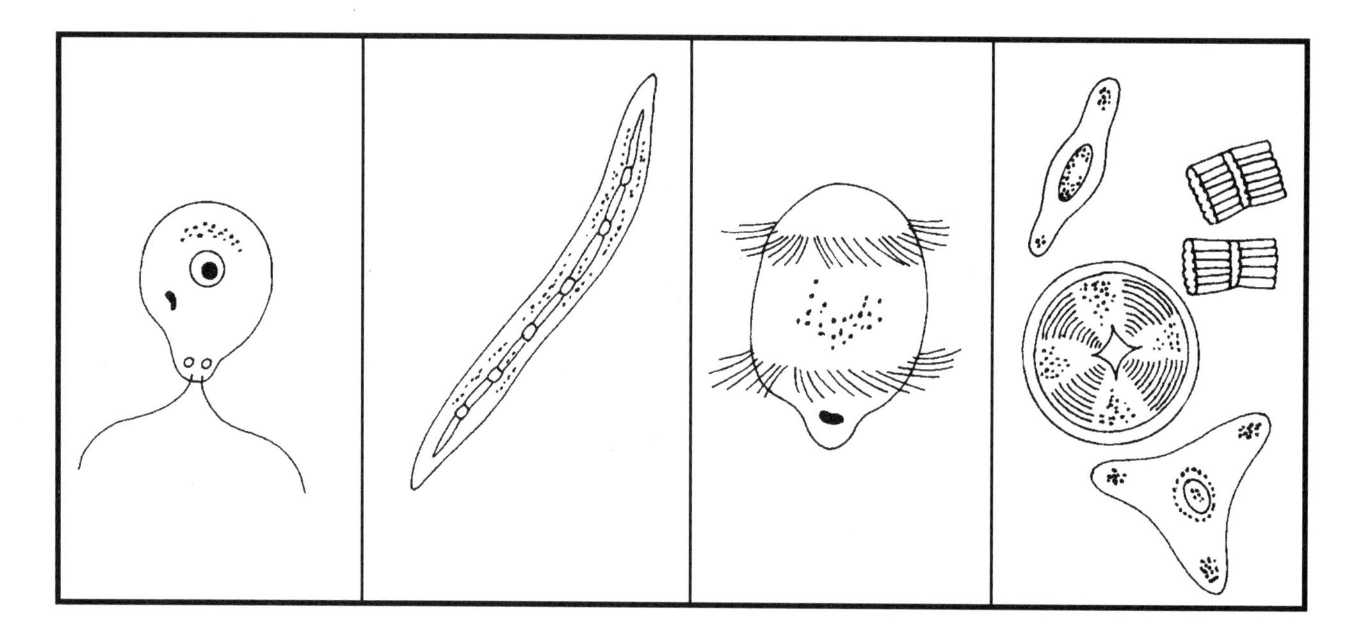

Go on to the next page.

Name ______________________________ Date ______________

MORE ONE-CELLED ORGANISMS, P. 2

Decide whether the statements below are true or false. Write a *T* on the line if the statement is true. Write an *F* on the line if the statement is false.

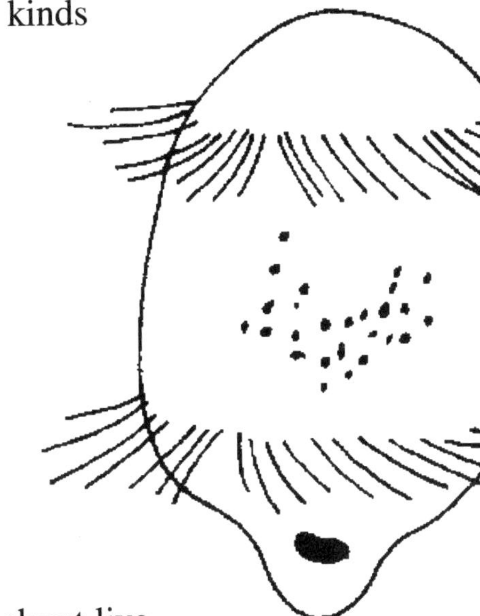

1. _____ There are thousands of different kinds of diatoms.

2. _____ Closterium are yellow-brown in color.

3. _____ Diatoms eat paramecia.

4. _____ Didinium have cilia.

5. _____ Chlamydomonas have 10 whiplike threads.

6. _____ All the organisms you just read about live in ponds.

Name ______________________________ Date ______________

Crunchy Celery

What gives celery its crunch? To find out, try this activity.

Materials:

- **glass**
- **water**
- **stalk of celery with leaves**

Do This:

1. Put a stalk of celery in an empty glass. Draw a picture in Box A to show what it looks like. Leave it overnight.

2. The next day look at your celery. Draw a picture of it in Box B.

   How does the celery feel? ________________
   Cut a small piece off the end. Bite into it.

   Is it crunchy? ________________

3. Add water to the glass. Set the celery aside for several hours. Then draw a picture of the celery in Box C.
   How does it feel? ________________
   Bite into it.
   Is it crunchy? ________________

   What made the difference? ________________

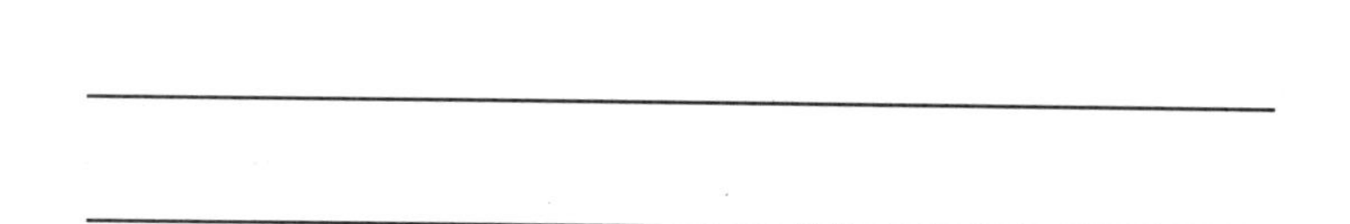

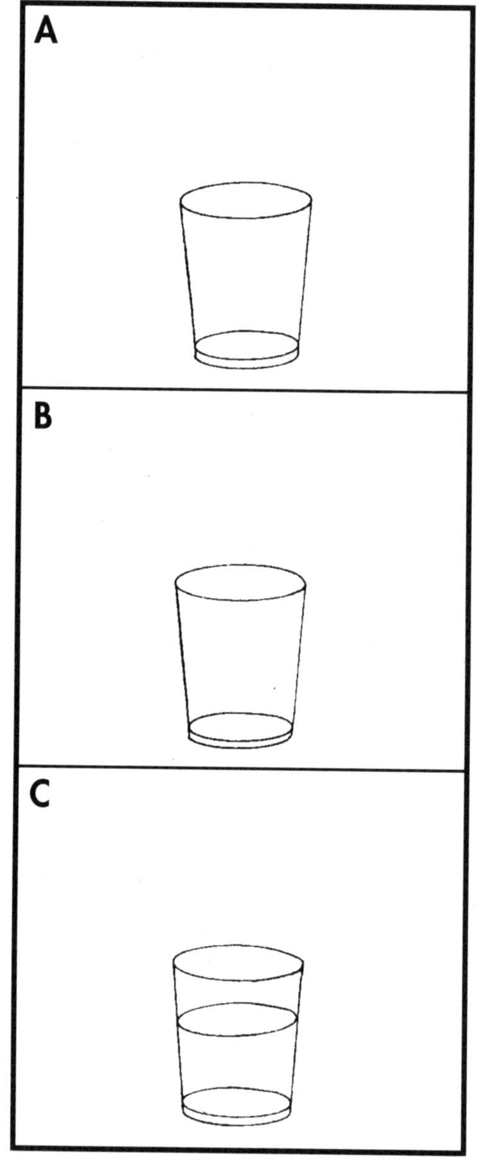

Celery is crunchy when its cells are stiff and full of water. When the cells lose water by evaporation, the celery becomes limp and droopy. When water diffuses back into the cells, the celery gets crunchy once again.

Name ______________________ Date ____________

Unit 1 Science Fair Ideas

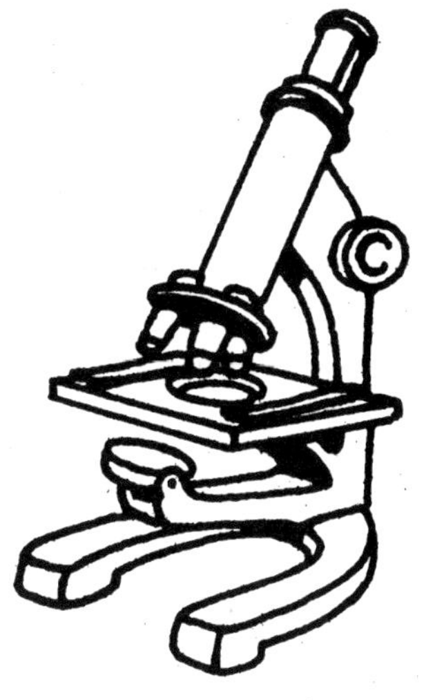

A science fair project can help you to understand the world around you better. Choose a topic that interests you. Then use the scientific method to develop your project. Here is an example:

1. **PROBLEM**: How do animal cells differ from plant cells?

2. **HYPOTHESIS**: Looking at animal cells and plant cells under a microscope can show the difference between the cells.

3. **EXPERIMENTATION**: Prepare slides with animal cells from places such as the inside of your cheek. Prepare slides with plant cells such as the cells in onion skin. Look at them under the microscope.

4. **OBSERVATION**: The plant cells and the animal cells do not look the same. The plant cells have cell walls.

5. **CONCLUSION**: Plant cells and animal cells are different. The differences can be seen under a microscope.

6. **COMPARISON**: Conclusion agrees with hypothesis.

7. **PRESENTATION**: Display your experimentation. Draw pictures of the cells that you saw and label their parts. Set up a microscope with prepared, labeled slides for people to look at. Help them to see the differences in the cells.

8. **RESOURCES**: Tell of any reading you did to help you with your experiment. Tell who helped you to get materials or set up your experiment.

Other Project Ideas

1. How do cells divide? What do the phases look like?
2. What does a plant need for photosynthesis to occur?
3. Do single-celled organisms exhibit all of the characteristics of living things?
4. Can you demonstrate the way a cell membrane works?
5. Why does a plant wilt when it needs water?

Unit 2: Plants and Animals
Background Information

Plant Classification

The plant kingdom contains about 450,000 different kinds of plants, which are classified into several divisions. The four main classifications for plants are:

- **algae** (almost all live in water; from microscopic single-celled plants to seaweed),
- **bryophyta** (mosses and liverworts; live in moist places; produce spores),
- **pteridophyta** (ferns, clubmosses, horsetails; no flowers), and
- **spermatophyta** (largest group—over 350,000 species; reproduce by way of seeds).

Spermatophytes are divided into two categories, the gymnosperms and the angiosperms. Gymnosperms, or "naked seed" plants, have seeds in cones, like pinecones from conifer trees. The angiosperms, or "covered seed" plants, include all of the flowering plants.

Flowering plants are the most numerous type of plant on Earth. They are further classified into groups. Some of the common groups of flowering plants are: grass family (corn, sugar, barley, rice, wheat), lily family (violets, hyacinths, tulips, onions, asparagus), palm family (coconuts, dates), rose family (strawberries, peaches, cherries, apples, and other fruits), legume family (peas, beans, peanuts), beech family, and composite family (sunflowers and others with flowers that are actually many small flowers).

Fungi are sometimes classified with plants, and sometimes not. A two-kingdom classification will include molds and fungi with plants, whereas a three-kingdom system will not. This is because molds and fungi lack chlorophyll and cannot produce their own food. They also lack roots, stems, and leaves, and they reproduce from spores. Mushrooms are also fungi.

Plants are also classified as vascular and nonvascular. Vascular plants have tubes that bring the liquids the plants need from their environment up through the plants. The tubes also help to support the plants. Nonvascular plants, such as mosses, do not have tubes. They are shorter because they must remain close to their source of moisture. They get the water and nutrients they need through their root systems.

Photosynthesis

Most plants are green. This is because they contain chlorophyll, most of which is in the leaves. Chlorophyll is contained in small structures in the leaf cells called *chloroplasts.* There are some plants that contain chlorophyll but whose leaves are not green. This is because the chlorophyll has been masked by other pigmentation in the plant. Chlorophyll is necessary for the making of food, but the chlorophyll itself is not used in the food that is made. Photosynthesis depends on light. A plant that is deprived of light loses its chlorophyll (and its ability to make food) and eventually will die. Plants take in the energy from the Sun and carbon, oxygen, and hydrogen from the air and water. Water and nutrients enter a plant through its roots. Carbon dioxide enters a plant through tiny holes (stomata) in the bottoms of the leaves. The plants change these raw materials into carbohydrates and oxygen. The carbohydrates (in the form of a simple sugar called *glucose*, and starch) are used and stored in the plants for food. The oxygen is released into the air and water where the plants live. In this way, plants constantly replenish the Earth's oxygen supply. Animals breathe the oxygen that plants supply. Animals also supply the carbon dioxide that plants need to survive. This is the oxygen-carbon dioxide cycle.

Green plants are the producers of a community. They not only produce their own food, but they also are the essential source of food and energy for all communities.

Plant Reproduction

Plants reproduce from seeds in flowers, from seeds in cones, or from spores. The seeds form after fertilization of their egg cells by male cells from pollen grains. Pollen can be carried to the egg cells by bees or other insects, by the wind, or by animals. Seeds contain tiny plants called *embryos*, around which a store of food is packed. In some seeds, such as bean seeds, the food is stored inside the embryo. Seeds are spread by animals and the wind. When the seeds in a cone are ripe, the cone will open and the seeds will float to the ground or be carried by the wind. Some seeds have tiny parachutes to help them drift. A seed needs moisture, warmth, and oxygen to begin growing into a new plant. If conditions are not right for germination, some seeds can remain in a resting state for hundreds of years.

Plant Life Cycles

The life cycle of a seed plant begins with an embryo. An embryo is an undeveloped living thing that comes from a fertilized egg. The eggs in a flowering plant are called *ovules*. When the ovules are fertilized, they begin to grow. A seed is a complete embryo plant surrounded by the food it needs to grow and protected by a coating. When the seed is planted, or lands on the ground, it begins to sprout. It grows into a seedling, then an adult plant that develops flowers in which new seeds grow.

Movement

When animal populations move, it is called *migration*. When plant populations move, it is called *succession*. Plants move for different reasons than animals do. If the vegetation in a particular area is destroyed, seeds will eventually move into the area. New life begins to grow. Grasses and small plants arrive first, then the pines, and finally hardwood trees.

Animal Classification

The animal kingdom can be classified into two large groups: the vertebrates (those with backbones) and the invertebrates (those without backbones). The backbone supports the body and provides flexibility. The spinal cord extends from the brain through the backbone, or spine. Individual nerves branch out from the spinal cord to different parts of the body. Messages from the brain are sent throughout the body through the spinal cord.

Some animals without backbones are sponges, jellyfish, clams, worms, insects, and spiders. Some of these animals have networks of nerves throughout their bodies with no central nerve cords. Many, like insects, have hard exoskeletons that protect their bodies and give them shape.

Mammals are identified as animals that have hair or fur, feed milk to their young, and are warm-blooded. Mammals are vertebrates. Amphibians are animals that live both on land and in water. Most lay eggs. Reptiles are scaly-skinned, cold-blooded animals. Birds are the only creatures with feathers. They are warm-blooded vertebrates. Fish are fitted to their environment because of their gills, which enable them to absorb oxygen from the water. Arthropods are animals without backbones. They have jointed legs, a segmented body, and an exoskeleton. Insects make up the majority of arthropods.

Animal Adaptations

In order for living things to remain alive, they must respond to changes or conditions in the environment. Environments include all the conditions in which a living thing exists. This includes the food the organism needs, water, soil, air, temperature, and climate. Common environments are deserts, grasslands, and forests. The living and nonliving things in each environment interact with each other to survive. When environments are threatened or changed in drastic ways, the living things in the environment are also threatened.

Animals are adapted to their environments through structures and behaviors. The structures include the physical makeup of animals. The behaviors of animals include things like migration and hibernation. Many of these behaviors are inborn. In winter, many birds migrate to warmer climates in the south. Some animals, like moose and caribou, also have migratory routes. Many animals hibernate, or sleep, through the winter months. They work through the fall to store food in their bodies that carries them through the winter months. While they sleep, their body processes slow.

Reproduction

All living things have a life cycle within which they take in food and gases, metabolize, excrete waste, reproduce, and die. If living things fail to repdroduce or to create healthy offspring, their species will die out. The California condor is one animal whose offspring have failed to thrive. So few of their chicks have survived in recent years that they may be in danger of extinction. It has been found that pesticides concentrated in the bodies of the adult condors have interfered with their reproductive abilities.

Animals reproduce in different ways. Some lay eggs, and others give birth to live young. Some offspring look like their parents, and others do not. Most reptiles, amphibians, fish, and insects lay eggs. The young of many of these animals can move about and find food for themselves soon after they hatch. Birds also lay eggs, but the adult birds remain with the eggs and care for the young until they can find their own food. Most mammals bear live young. The young are fed milk from the mother's body. Mammals spend more time than other animals feeding, protecting, and teaching their young to survive on their own. Animals that give birth to live young have fewer offspring than those that do not tend to their young. The young of human beings require more care from their parents than any other animal.

Metamorphosis

The life cycles of some animals include a metamorphosis. A metamorphosis is a complete change in the appearance of an animal. The most striking metamorphosis is the change from caterpillar to butterfly. Metamorphosis is controlled by hormones in the body. When the hormone supply keeping a caterpillar a juvenile stops, the caterpillar begins to become a chrysalis, or pupa. In a frog, the change is controlled by the thyroid gland. Crabs also undergo metamorphosis, and earwigs and grasshoppers undergo incomplete metamorphosis.

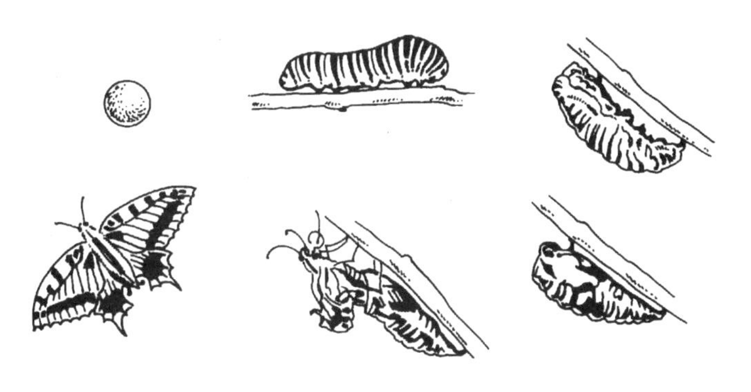

The Biosphere

The biosphere includes the atmosphere, the upper surface of the Earth's crust, and the oceans. The Sun can also be considered part of the biosphere, as its energy is used by living things. Life on Earth is contained in the biosphere. Here living things grow, reproduce, and die. In the process, they interact with each other, with other living and nonliving things, and with their environment. They change to adapt to their environments, and they change their environments. Any study of the biosphere includes the study of the relationships between the plants and animals that live there. The interactions between the plants and animals in the biosphere consist of energy chains, or food chains.

The Web of Life: Life Cycles, Communities, and Food Chains

All living things go through life cycles. From single-celled organisms to the largest animals, these life cycles include growth, change, consumption of food and water, use of energy, reproduction, and death. Reproduction varies among life forms. Plants reproduce by seeds or spores. Animals may lay eggs or give birth to live young. Some offspring resemble the parents, and others do not. Some animals, such as frogs, undergo metamorphosis, or a complete change, during their lifetimes. The successful reproduction of a species is important to that population's continued growth or stability.

Populations are plants or animals of one kind that live in one area. Populations interact to form communities. Each community may have many different habitats. Each population has its own habitat. The interactions of populations in a community create food chains. These interdependencies are known as *symbiosis*, or living together in close association.

A typical food chain begins with plants. Most plants make their own food. Then they pass off oxygen to be used by other living things. Plants also produce sugar and starch, which are used by other animals. The animals that use plants are herbivores (plant eaters), carnivores (meat eaters who eat the plant eaters), and omnivores (plant and animal eaters). Animals give off carbon dioxide, which is used by the plants. Food webs are used to describe overlapping food chains. Communities and the interactions within them are complex. When their natural order is disrupted, the balance of nature is affected, and organisms can be in danger. The most dire consequence of this disruption is the extinction of a species.

The relationships between organisms in a community can be described in three ways. If the relationship between two organisms is beneficial to both, it is called *mutualism.* If the relationship helps one organism while the other is neither helped nor harmed, it is called *commensalism.* If the relationship helps one organism and harms the other, it is known as *parasitism.*

A community contains food makers, or producers, food takers, or consumers, and decomposers. Typically, the producers are plants. Both other plants and animals eat plants. Carnivores also need plants as they live off the animals that eat plants. Consumers eat plants, animals, or plants and animals. An animal that eats another animal is a predator. The animal that is eaten is the prey. When the producers and consumers die, they begin to change—they rot and decay. The decomposers get their food from wastes and dead organisms. Molds, yeast, and bacteria break down the dead matter and give off carbon dioxide. The carbon dioxide is then used by green plants to make food.

Biomes

A biome is a large community of plants and animals. Biomes are characterized by the plants and animals that are found there and by the biome's specific climate. There are six major land biomes on Earth.

- The tropical rain forest is warm and rainy. Its animal life is primarily in the trees.
- The deciduous rain forests have warm to hot summers and cold winters. The trees there lose their leaves in winter.
- The boreal forests are very cold and snowy in winter and have very short growing seasons. The trees there are primarily evergreens.
- The arctic tundra has long, cold winters and cool, short summers. There is not much precipitation.
- The grasslands have hot, dry summers and cold, snowy winters.
- The deserts receive very little precipitation. Most are hot year-round, but the nights can be cool because there are no clouds to hold in the warm air.

The Human Factor

Organisms need to adapt to and change with the changes in their environment to survive. If they cannot adapt, they will not survive. Organisms adapt through physical changes that help them live in their particular habitats and through habits, such as migration, that help them survive. Although many events can disrupt a community and its balance, humans have had the greatest impact upon the Earth's environment. Humans need not only food and energy but also power and space for settlement. Humans create wastes that are not natural to the environment. This environmental pollution is an important concern for everyone. If it is not controlled, the balance of nature is disrupted, organisms die, and those that depend upon the dead organisms may die. It is crucial for humans to find ways to live without creating such disturbances in the environment.

Name ______________________ Date ____________

CLASSIFYING PLANTS

A. Green plants provide the air with something you need—oxygen. They also use the carbon dioxide that you exhale when they produce food. This relationship between you and plants is called the oxygen-carbon dioxide cycle. Below are the steps of this cycle. In each step, the words have been scrambled. Put the words in their proper order.

1. in oxygen animals breathe

____

2. breathe out dioxide carbon animals

____

3. carbon dioxide in plants take

____

4. own chlorophyll make food and their plants use carbon dioxide water sunlight to

____

____

5. off oxygen plants give

____

B. Scientists sometimes summarize their observations by making a diagram. In the space below, make a diagram that shows the five steps in the oxygen-carbon dioxide cycle.

Name ______________________________ Date ______________

ROOTS, STEMS, AND LEAVES

Would you like to take an incredible journey? Here is one for you. Imagine that you are a drop of water in the ground near a plant.

The following is a description of your travels, but it is missing some important words. Write in words from the box that will make this journey complete. You may use a word more than once.

chlorophyll
food
leaves
minerals
photosynthesis
stem
sunlight
support
taproot
water

I am deep under the ground where it is very dark. I am near an alfalfa plant. I am slowly moving toward a very large root. It must be the ____________________. I am now entering the root. Beside me are ____________________ that will eventually be used by the plant. I am moving upward through a system of tubes. I must be in the ____________________. The stem carries ____________________ and ____________________ upward. It also gives the plant ____________________. Upward and upward I travel. I am now moving into an area that seems very green. I can see the cells of the ____________________ straight ahead. There is green material in them. It must be ____________________. I am now passing through the cell wall and am in the cell. I feel a lot of energy coming into the cell. The energy must be ____________________. Carbon dioxide has also entered the cells. We are about to be changed into ____________________. The gas oxygen is leaving. The process of ____________________ has taken place. The journey is over.

Name ______________________ Date ____________

How Do Substances Move up the Stems of Plants?

To find out, try this activity.

Materials:
- food coloring
- scissors
- glass filled $\frac{1}{4}$ with water
- hand lens
- long-stemmed white flower

Do This:
1. Add a few drops of the food coloring to the glass of water.
2. Trim the stem of the long-stemmed white flower.
3. Put the flower in the glass of water and leave it overnight.

Answer these questions.
1. What does the flower look like the next day?

   ____

2. What happened to make the flower look this way?

   ____

3. Use the scissors to cut the stem above the waterline. Use the hand lens to look at the cut top of the stem in the glass. What do you observe?

   ____

Draw a picture of what the cut top of the stem looked like through the hand lens. Write a paragraph explaining how you think water is transported up through a stem.

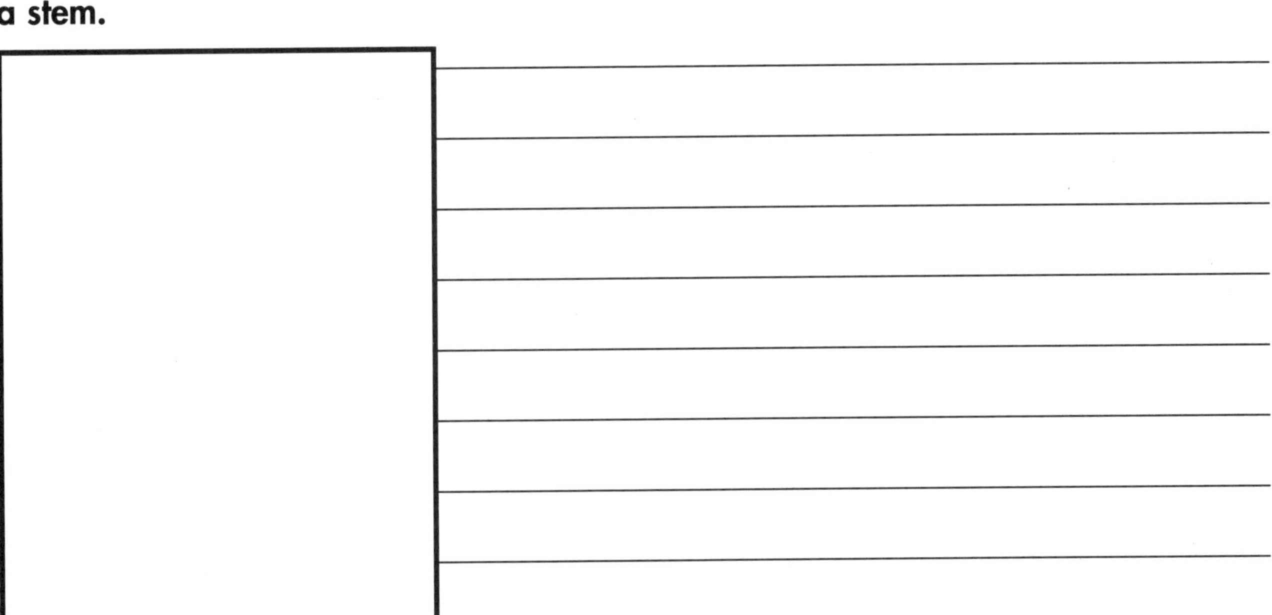

Name ______________________ Date ____________

DO PLANTS LOSE WATER?

You are learning about how water moves through plants. In this activity, you will observe how some of the water leaves the plant.

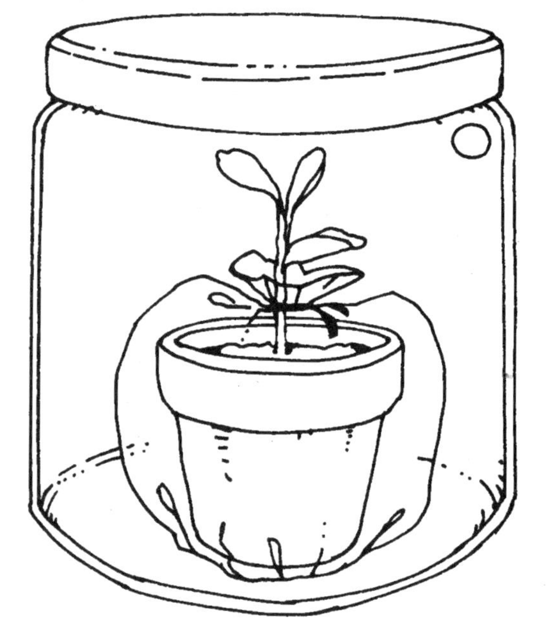

Materials:

- **small potted plant**
- **string**
- **plastic bag**
- **large jar with a lid**

Do This:

1. Water the plant so that the soil is moist.
2. Place the pot into the plastic bag and tie the bag to the bottom of the stem.
3. Place the plant inside the jar and screw the lid on tightly.
4. Place the jar in a sunny spot for a few hours. Observe what happens inside the jar.

Answer these questions.

1. What happened in the jar? ______________________

____

____

____

2. Why did this happen? ______________________

____

____

____

Name ______________________ Date ____________

WHAT HAPPENS TO A PLANT WITHOUT SUNLIGHT?

To find out, try this activity.

Materials:

- **two green plants of the same type and size**
- **box**
- **water**

Do This:

1. Examine the two similar green plants. Notice the color of their leaves. Be sure the soil in each pot is moist. Label the plants **a** and **b**.
2. Place plant **a** on a sunny windowsill. Place plant **b** on a table, and put a box over it.
3. Water the plants as needed.
4. Observe the plants one week later.
5. Place plant **b** in the Sun for one week. Observe it.

Answer these questions.

1. After the first week, what color were the leaves of plant **b**? ____________
2. How did they compare to the leaves of plant **a**? ____________

   ____
3. After being moved into the sunlight for a week, what color were the leaves of plant **b**?

   ____
4. What was missing from the leaves of plant **b** after the first week in the dark? ____________

   How do you know? ____________
5. What caused plant **b** to change after a week in the Sun? ____________

   ____
6. What is the relationship between sunlight and the amount of chlorophyll in a plant?

   ____

   ____

Name ______________________ Date ______________

Plant Reproduction

Reproduction is the method by which living things make more of their own kind. There are several ways that plants reproduce.

- *Mosses* are small plants with no real roots, leaves, or stems. Growing in moist, shady places, mosses get the water and nutrients they need by living close to the soil and passing the food and water slowly from cell to cell. Mosses reproduce by spores. Spores are special cells that can develop into new plants like the plants that made them.

- *Ferns* also grow in moist, shady places. Ferns have roots, stems, and leaves, however, and can grow very large. Ferns also reproduce by spores. The spores grow on the underside of the fern leaves. When the tiny pockets of spores break open, thousands of spores float away on the wind to begin new ferns.

- *Gymnosperms* are seed-bearing plants that produce seeds in cones. Fir trees are common gymnosperms. When the seeds are ready, the cones release them into the wind. When the seeds come to rest in the right kind of soil, they will begin to grow a new plant.

- *Angiosperms* are seed-bearing plants that are also called flowering plants. They are the most common type of plant. All flowering plants and trees are angiosperms. Grass, flowers, and fruit trees all produce seeds. Each seed contains a tiny plant and stored food to help it grow.

Answer these questions.

1. What kind of plant is a pine tree? ______________________

2. How would a pine tree reproduce? ______________________

3. What is the most common type of plant? ______________________

4. What is in a flower seed? ______________________

5. What is a spore? ______________________

Name ______________________ Date ____________

How Plants Reproduce

Label the two plant cycles shown below. Then cut out the stages of each. Arrange them in the proper order and paste them onto a sheet of paper.

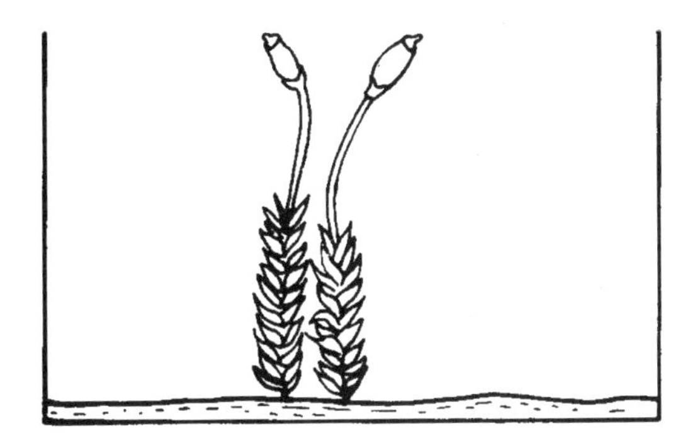

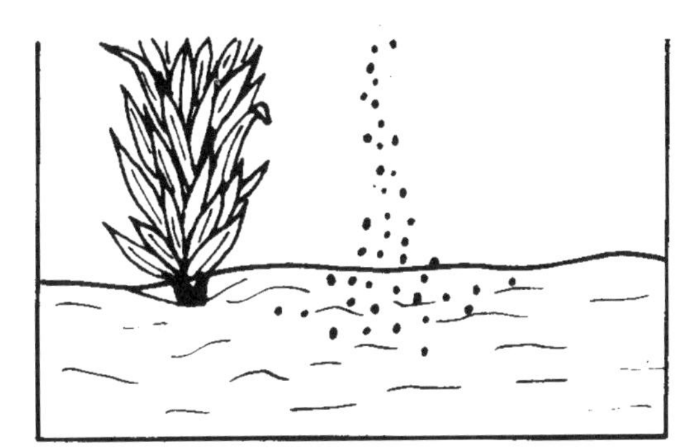

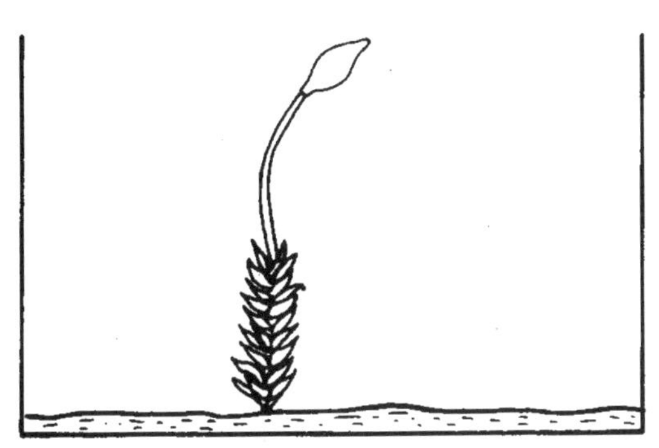

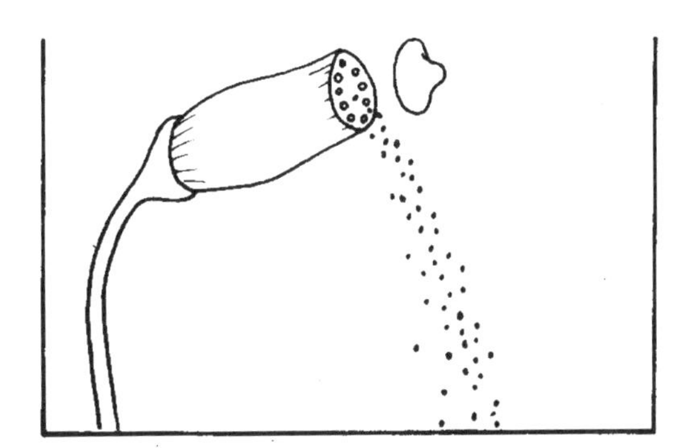

Reproduction by ______________________

____

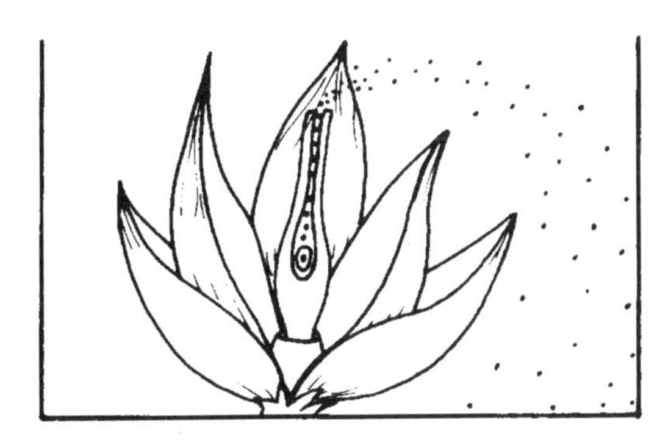

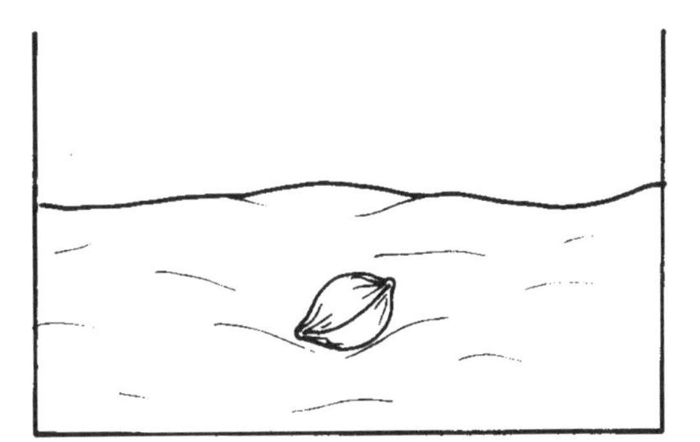

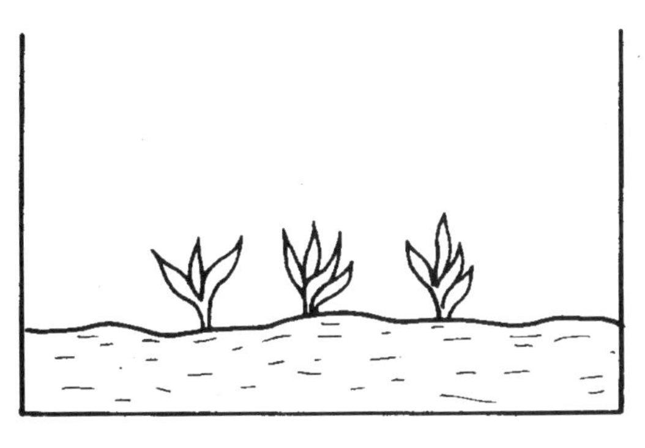

Reproduction by ______________________

Name ______________________ Date ____________

Groups of Plants

Look at the plants in the picture below. Then write your answer to the questions.

1. Classify each of the plants you see in the picture into the groups listed in the margin. Then describe how each type of plant reproduces.

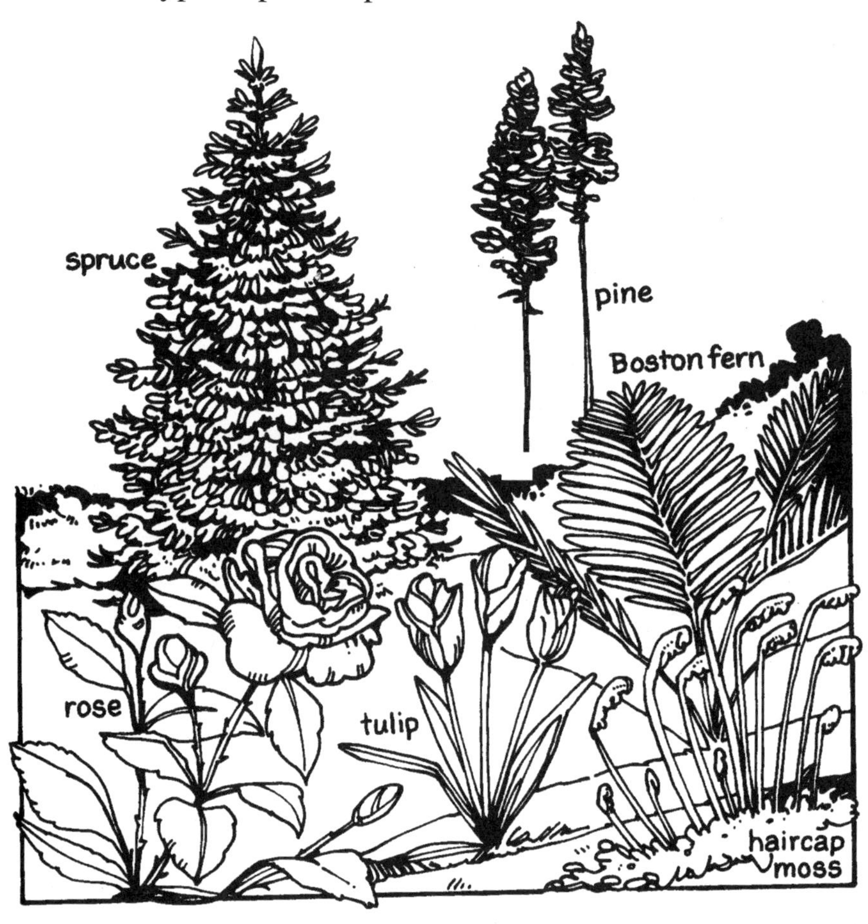

2. Write a hypothesis that explains why the haircap moss will not grow as tall as the spruce tree.

____

____

____

____

____

____

Classification

Angiosperms

____

____

Gymnosperms

____

____

Ferns

____

____

Mosses

____

____

Reproduction

Angiosperms

____

____

Gymnosperms

____

____

Ferns

____

____

Mosses

____

____

Name ______________________ Date ______________

Food for Growth

You need to eat every day. Food gives you energy. It also provides you with the materials you need to grow and repair body tissue.

Plants need food for the same reasons. You can see how a bean seed uses some of its stored food.

Materials:
- **four bean seeds**
- **water**
- **soil**
- **paper cup (plastic-coated)**
- **aluminum foil**

Do This:

1. Soak the bean seeds in water overnight. Then look at them. Describe them.

   ____

How do they feel? ______________________________

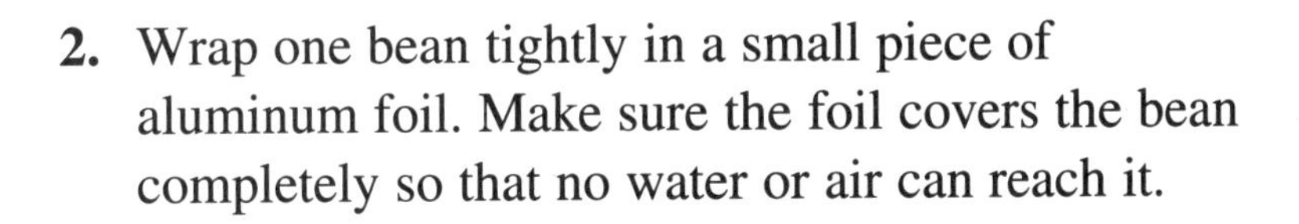

2. Wrap one bean tightly in a small piece of aluminum foil. Make sure the foil covers the bean completely so that no water or air can reach it.

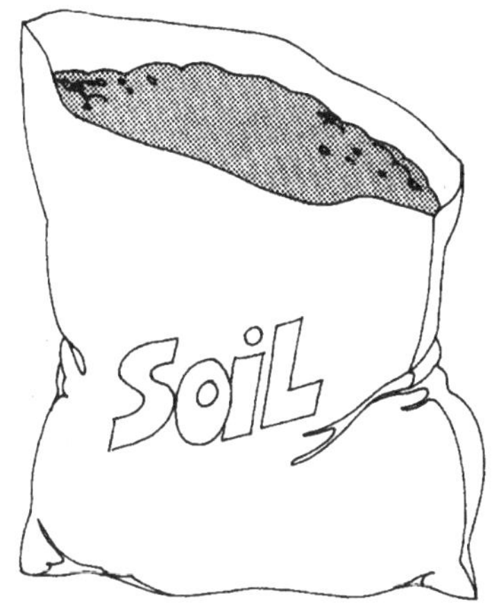

3. Fill your paper cup $^3/_4$ full of soil. Plant the beans about 3 cm deep in the soil. Water the soil. Keep it moist for one week.

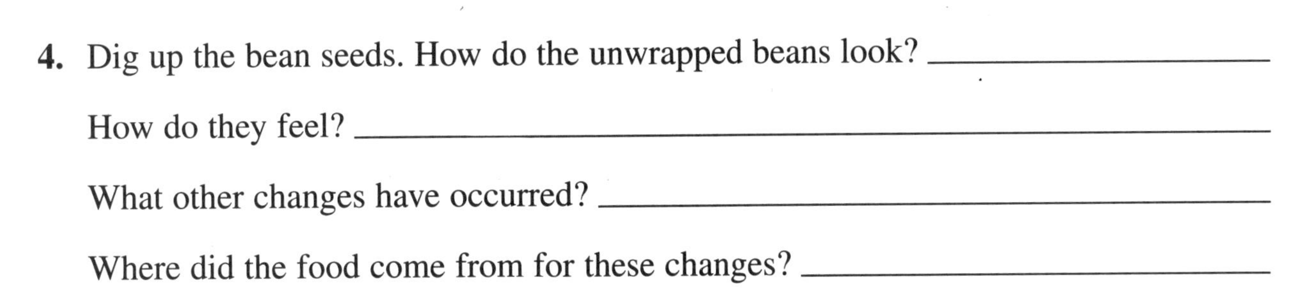

4. Dig up the bean seeds. How do the unwrapped beans look? ______________

   How do they feel? ______________________________

   What other changes have occurred? ______________________________

   Where did the food come from for these changes? ______________________________

5. Unwrap the bean in aluminum foil. How is it different from the other beans?

   ____

With the right conditions, a bean seed will use its stored food to grow a new plant. If the bean has no oxygen or water, however, it cannot use the stored food.

Name ______________________ Date ____________

FOOD CHAINS

There are two types of plankton—animal and plant. Plant plankton captures light energy from the Sun and coverts it to food. Most animals in the sea, including animal plankton, ultimately depend upon plant plankton for food. All these tiny plants use only a half percent of the light energy that strikes the surface of the ocean. Yet they help feed an ocean full of animals!

A food chain shows how animals depend on other organisms for food. Most ocean food chains begin with plant plankton. Most land food chains begin with plants, also.

Study the following food chain.

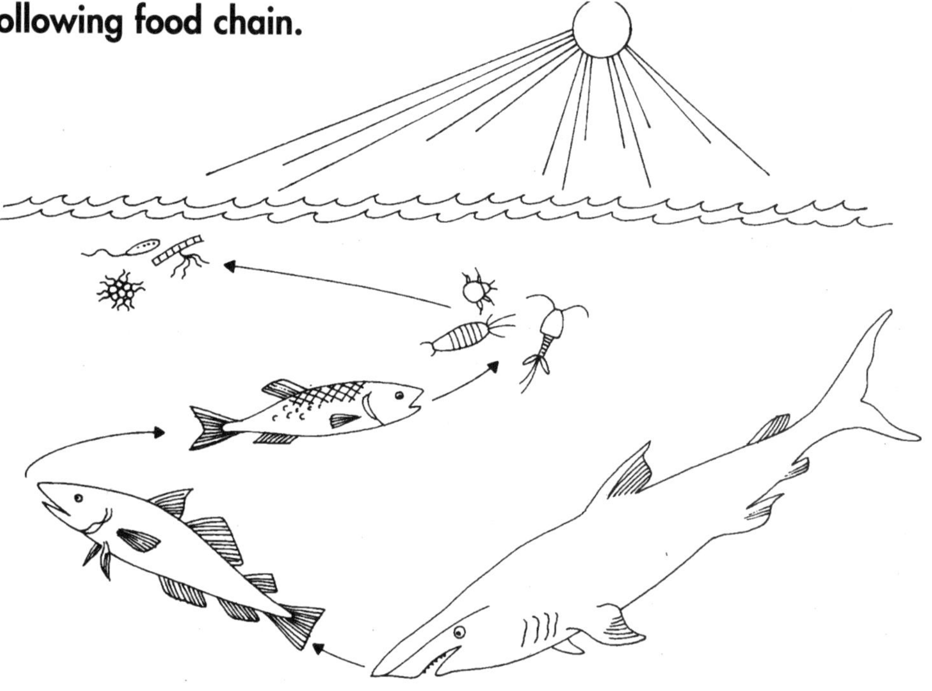

Answer these questions.

1. What does the cod eat? ______________________

2. What eats the cod? ______________________

3. Where does the energy for this food chain come from? ______________________

4. What is the smallest organism in this food chain? ______________________

5. What would happen if all plant plankton suddenly disappeared? ______________________

____

____

Name ______________________________ Date ______________

ENDANGERED PLANTS

Read the following story about endangered plants. Then complete each sentence on the next page by circling the best answer.

Like some animals, plants can become extinct, too. This happens when all the members of a plant group die off and leave none to reproduce. About 100 kinds of plants that once grew in the United States are now extinct in the wild. Hundreds of other plants are in danger of dying off.

One endangered plant is the giant sequoia tree. Giant sequoias may be the oldest of all living things. Some may live for more than 3,000 years. Sequoias may be the biggest living things, too. The largest is the "General Sherman," which is 83.8 meters high. Giant sequoias grow only in certain parts of California. National parks have been created in these areas to help save these endangered plants.

Several kinds of pitcher plants are also endangered. Pitcher plants have leaves shaped like pitchers. When it rains, water collects in the leaves. Insects drown in the rain water and become food for these animal-eating plants. Pitcher plants need to live in wet places to survive. Many wetlands, however, disappear as people use land for farms, highways, and homes. The disappearance of the wetlands means fewer places where pitcher plants can grow.

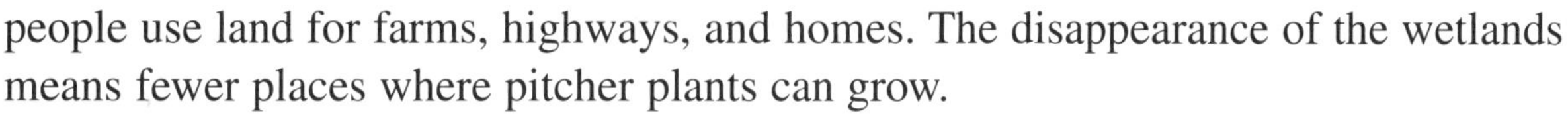

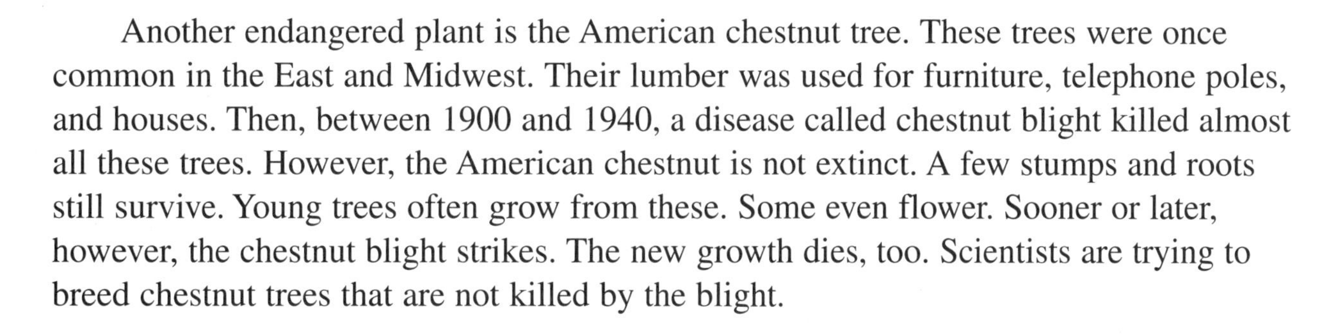

Another endangered plant is the American chestnut tree. These trees were once common in the East and Midwest. Their lumber was used for furniture, telephone poles, and houses. Then, between 1900 and 1940, a disease called chestnut blight killed almost all these trees. However, the American chestnut is not extinct. A few stumps and roots still survive. Young trees often grow from these. Some even flower. Sooner or later, however, the chestnut blight strikes. The new growth dies, too. Scientists are trying to breed chestnut trees that are not killed by the blight.

Go on to the next page.

Name ______________________ Date ______________

ENDANGERED PLANTS, P. 2

1. Plants and animals become extinct if none of the members of the group (travels reproduces).

2. About (100 1,000) kinds of American plants are extinct in the wild.

3. Giant sequoias are some of the (smallest largest) living things.

4. Giant sequoias live only in certain parts of (Colorado California).

5. Pitcher plants live in (wet dry) places.

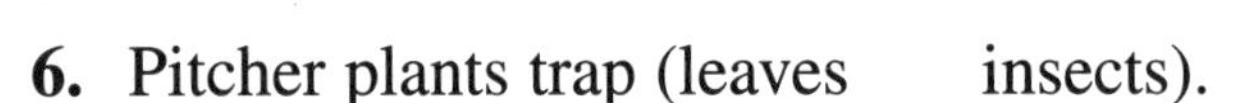

6. Pitcher plants trap (leaves insects).

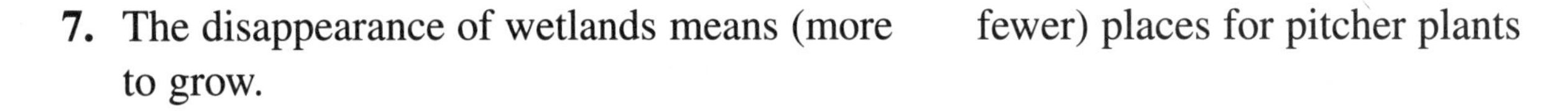

7. The disappearance of wetlands means (more fewer) places for pitcher plants to grow.

8. Most American chestnut trees were destroyed by (blight lumbering).

Name ______________________ Date ____________

BACKBONE OR NO BACKBONE

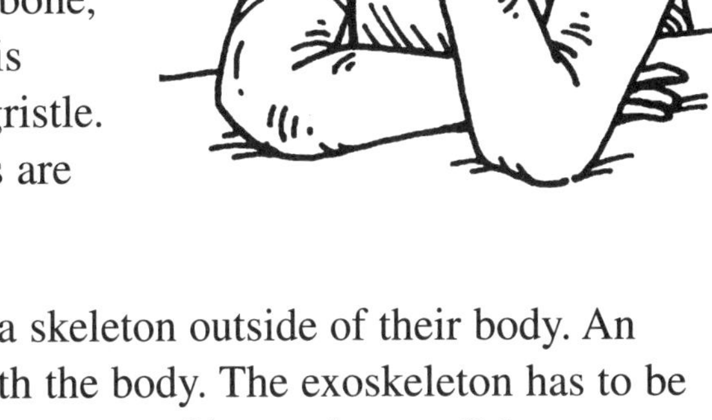

Animals are often divided into two categories—those with backbones and those without backbones. Animals with backbones are called *vertebrates*. Animals without backbones are called *invertebrates*.

Vertebrates have an internal skeleton, or a skeleton inside the body. The skeleton is usually made of bone, but in some cases, as in the shark, the backbone is made of cartilage. Cartilage is a tough, rubbery gristle. Fishes, amphibians, reptiles, birds, and mammals are all vertebrates.

Many invertebrates have an exoskeleton, or a skeleton outside of their body. An exoskeleton is very strong, but it cannot grow with the body. The exoskeleton has to be shed when the body grows. Insects are the largest groups of invertebrates. Other invertebrates are snails, jellyfish, worms, sponges, crabs, and lobsters.

Write the names from the box under the correct heading.

Invertebrates	Vertebrates
____________	____________
____________	____________
____________	____________
____________	____________
____________	____________
____________	____________
____________	____________
____________	____________
____________	____________
____________	____________
____________	____________
____________	____________

starfish	ant
turtle	frog
lobster	horse
spider	mouse
jellyfish	dog
worm	bear
sponge	eagle
beetle	fish
lizard	bee
snail	crab
shrimp	rabbit
human	snake

Name ______________________ Date ____________

KINDS OF ANIMALS

Mammals are covered in fur or hair. They give birth to live young and feed them milk from their body. They are warm-blooded. This means that they can maintain a warm body temperature in cold weather. Mammals all have backbones, and they can be found all over the Earth.

Amphibians are animals that can live both on land and in water. Most breed in the water and lay eggs. The eggs develop into tadpoles and then adults. Amphibians can be found everywhere except Antarctica and Greenland.

Reptiles are scaly-skinned animals. Some live in water, and some live on land. Reptiles are cold-blooded. They control their body temperature by the Sun. If they need heat, they sit in the Sun. If they are too warm, they move to the shade. Most reptiles are found in warm areas of the world.

Birds are the only creatures with feathers. They are warm-blooded animals. Most birds can fly, but some, such as the ostrich, cannot.

Fish are streamlined for swimming. They have a strong tail, and fins for balance and steering. Fish absorb oxygen from the water with their gills.

Arthropods are animals without backbones. They have jointed legs, a segmented body, and an exoskeleton. Insects are the largest group of arthropods.

Use a science book, an encyclopedia, or what you already know to find pictures of each type of animal. Write the following headings on a piece of poster board, and paste your pictures under the correct headings.

Mammals **Amphibians** **Reptiles** **Birds** **Fish** **Arthropods**

Name ______________________ Date ______________

SIMPLE ANIMALS

This exercise reviews some characteristics of simple animals. You may use a science book or encyclopedia to help you find the answers.

A. Imagine that you have just discovered an animal at the bottom of a pond. You name it the Mudwumple. You must decide if it is a vertebrate animal or an invertebrate animal.

1. What major body structure would you look for on the Mudwumple? Why?

____

____

____

2. The Mudwumple is an invertebrate. You think that it is related to the sponge. What structures would the Mudwumple have if it were related to the sponge?

____

____

____

3. Now you notice something else. Around its opening are small tentacles. What group of animals might the Mudwumple be related to now?

____

B. Below are drawings of three different types of worms. Under each, write the name of the type of worm it is. Also describe its characteristics.

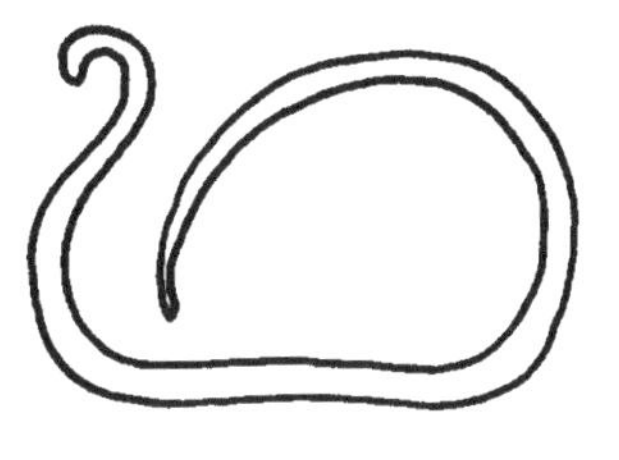

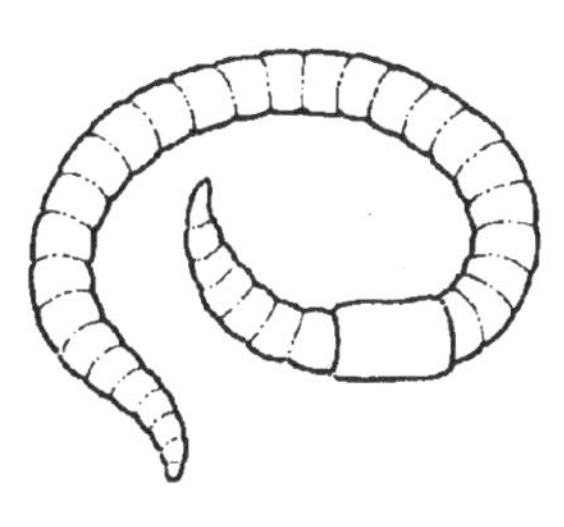

______________ ______________ ______________

______________ ______________ ______________

______________ ______________ ______________

Name ______________________________ Date ______________

Animals with Shells and Spines

Complete these exercises.

A. This aquarium is overflowing with sea animals. Some are mollusks. The rest are spiny-skinned invertebrates. Separate the two groups of names into the smaller aquariums below. Place all the mollusks together and all the spiny-skinned invertebrates together.

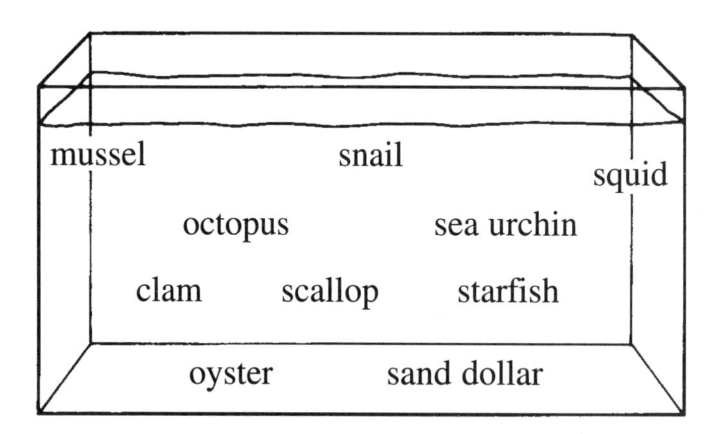

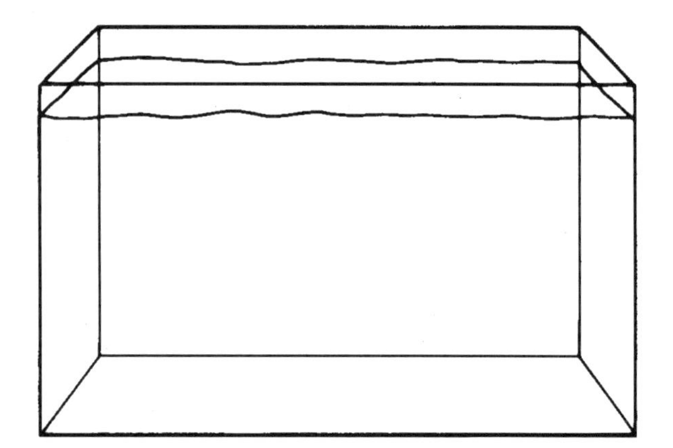

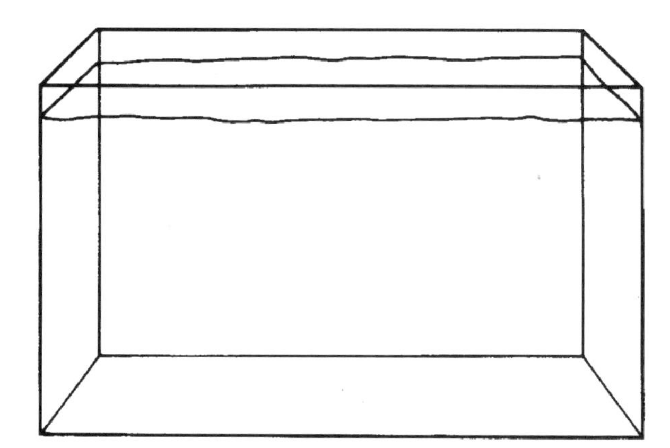

B. Starfish feed on mussels. Make a series of drawings below that show how a starfish grasps and eats mussels. Under the drawings, explain the steps that you have shown.

Name ______________________ Date ______________

How Does a Snail React?

To find out, try this activity.

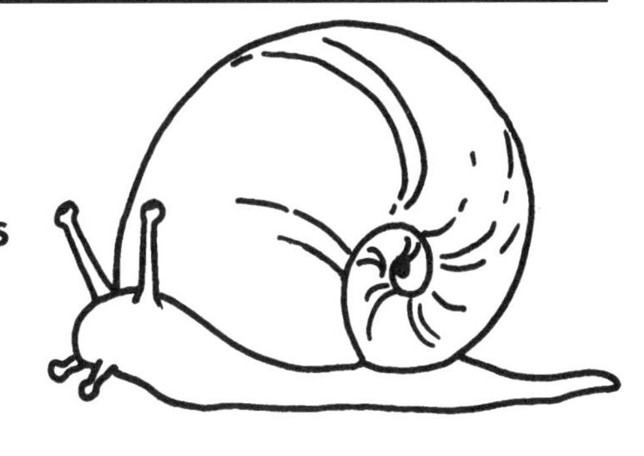

Materials:
- **aquarium with freshwater snail**
- **hand lens**

Do This:

1. Observe the snail crawling up the glass side of the aquarium. Use a hand lens to see how the snail's "foot" moves. How does the foot move?

   ____

   ____

2. The front part of the foot is the snail's head. Look at it with the hand lens while the snail is moving. What structures can you see on the snail's head?

   ____

   What do you think these structures do? ______________________

3. Tap the aquarium sharply near the snail. What happens? ______________________

   Does the snail react quickly? ______________________

Answer this question.

4. Write a paragraph describing how the snail reacted when you tapped on the glass. Tell whether or not you think this reaction is a useful one, and why you think so.

   ____

   ____

   ____

   ____

   ____

   ____

   ____

Name ______________________ Date ______________

Collect, Study, and Release Insects

Arthropods are animals without backbones. They have jointed legs, a segmented body, and exoskeleton. Insects are the most numerous of all arthropods. In this activity, you will study some of them.

Materials:

- **three glass jars with lids**
- **insect net**
- **insect reference book**
- **hand lens**

(If you don't have an insect net, you can make one by attaching the toe of a nylon stocking to a hoop made from a coat hanger.)

Do This:

1. Capture three insects using the insect net. Do not touch the insects. Place one in each of the glass jars.
2. Use the hand lens to observe each insect.

Answer these questions.

1. What are the names of the insects you have captured? Ask your teacher for help, or look the names up in a reference book. ______________________
2. Make a chart for each insect. On it, describe the size and color of each insect.
3. Make a drawing of each insect. Use the reference book to find out what the insect feeds on.
   Release each insect outdoors when you are finished.

monarch butterfly

grasshopper

ant

bee

Name ______________________________ Date ______________

OBSERVING CRICKETS

Have you ever listened to the chirp of crickets? In this activity, you will raise some crickets of your own.

Do This:

1. You will need five to ten male or female crickets. Catch them yourself or buy them in a fish bait store. You will also need a large plastic container and a cover with holes, a smaller plastic container with a cover, enough soil to fill the bottom of the large container to a depth of 5 cm (about 2 inches), a small bowl or saucer, pieces of cardboard, water, dry breakfast cereal, and food such as a potato or carrot.

2. Prepare a home for the crickets. Put the soil in the bottom of the large container. Bury the bowl at one end of the container so that the top edge of the bowl is level with the soil's surface. Fill the bowl with water and keep it filled. Crumple the cardboard and put it in the container away from the water.

3. Place some dry cereal and a few slices of carrot or potato in the large container. Place the crickets in the container and put on the cover. Observe them. Are they chirping?

4. Remove one cricket and place it in the small container. Put the cover on the container and put the container in a warm place for an hour. How many chirps does it make in a minute? Write down your observation.

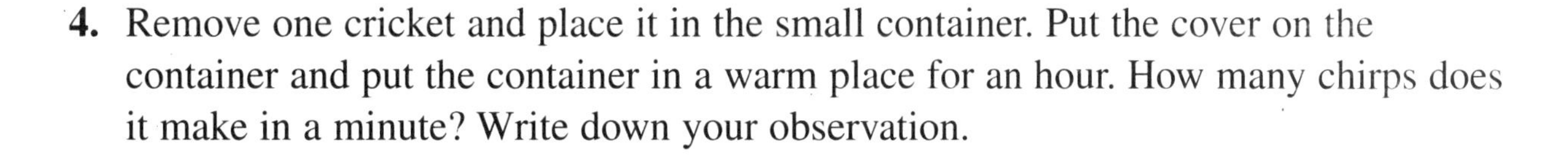

____

5. Put the cricket in a cool place for an hour. How many chirps does it make in a minute now?

____

6. The number of chirps a cricket makes has to do with the temperature of the cricket's environment. Can you predict what will happen to the number of chirps the cricket will make at an even higher temperature?

____

Name ______________________ Date ______________

INSECT DEVELOPMENT

One group of arthropods is far more numerous than all the others.

Answer these questions.

1. What is the name of this group? ______________________

2. Make a hypothesis that explains why there are so many of these arthropods.

____

____

____

____

3. Arthropods sometimes develop in stages. Below is a diagram that shows these stages. Write in the missing labels from the box.

• butterfly egg stage	• caterpillar	• pupa
• adult butterfly	• butterfly shedding its covering	

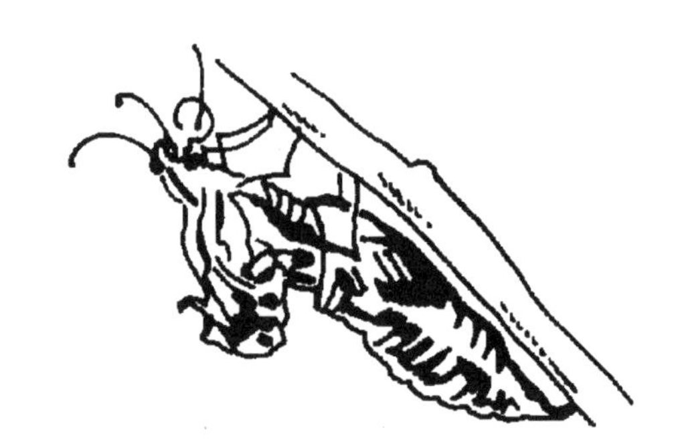

Name ______________________________ Date ______________

Caterpillar Metamorphosis

You have been studying the metamorphosis of insects. In this activity, you will observe the changes a caterpillar undergoes.

Materials:

- **a large glass jar with a lid**
- **a plastic bag**
- **fresh leaves and twigs**

NOTE: An adult should help you punch holes in the lid of the jar.

Do This:

1. Find a caterpillar eating a leaf. Place it into a plastic bag along with some leaves and twigs from the plant.
2. Carefully place the contents of the plastic bag into the jar.
3. Observe how the caterpillar moves and eats. If possible, watch it spin its cocoon.

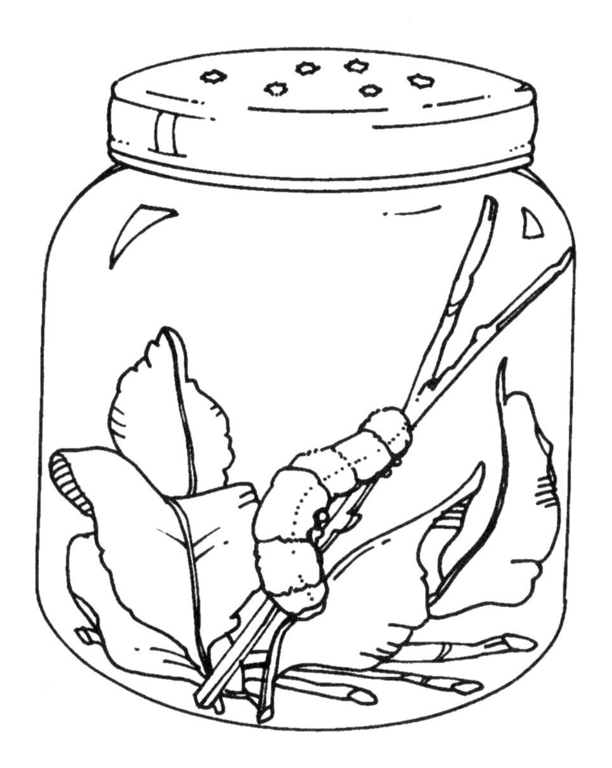

Answer these questions.

1. How many legs does the caterpillar have? ______________________________

2. What is happening when the caterpillar is inside its cocoon?

____

____

____

____

____

Name ______________________________ Date ________________

Classifying Invertebrates

Sometimes students make collections of invertebrate animals. Here is a way of collecting invertebrates without even going outside.

The names for groups of invertebrates are shown below. Under them are three nets. Write the name of each group in the proper net.

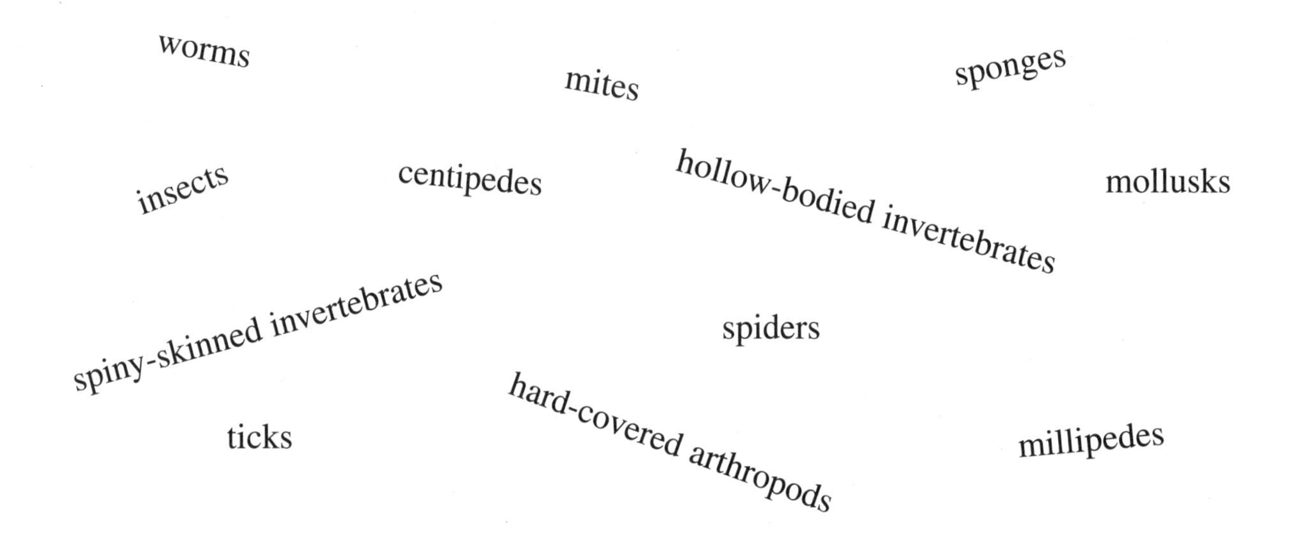

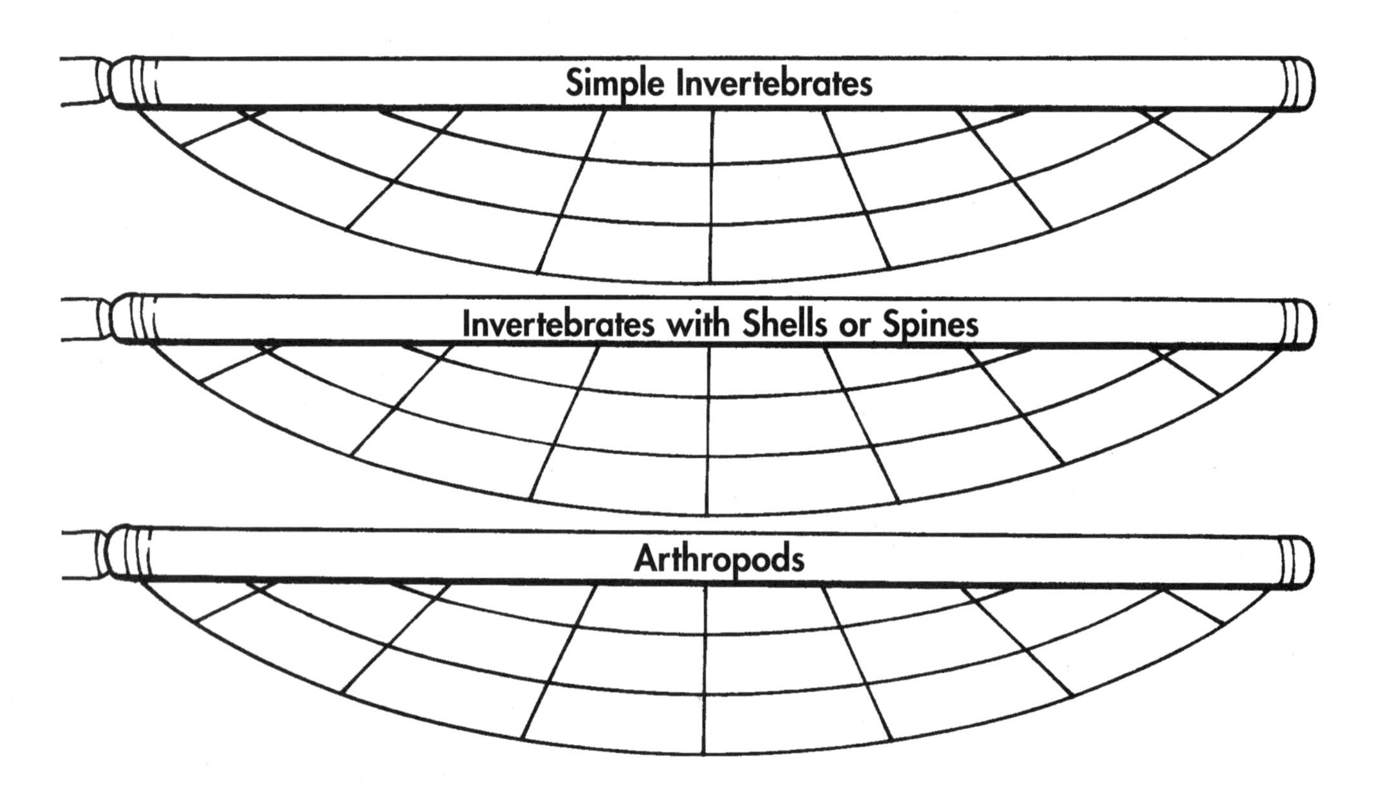

Name ______________________________ Date ______________

Who Am I?

Have you ever played the game "Who Am I?" Read each question below. Then guess the animal by using what you have learned. Some answers have more than one animal.

______________ 1. I am a mollusk with no shell and a "foot" with eight or more tentacles. Who am I?

______________ 2. I am a simple invertebrate with one large body opening and smaller holes over the rest of my body. Who am I?

______________ 3. I am a worm with a body divided into many similar body parts. What kind of worm am I?

______________ 4. I am a kind of animal that has no backbone. What kind of animal am I?

______________ 5. I am a mollusk with no hard shell, but I am jet-powered. Who am I?

______________ 6. I am a worm with one body opening and a long flattened body. What kind of worm am I?

______________ 7. I am a kind of animal that has a backbone. What kind of animal am I?

______________ 8. I am a spiny-skinned invertebrate with five arms that are covered with tube feet. Who am I?

______________ 9. I am an invertebrate animal with jointed legs and an exoskeleton. What kind of animal am I?

______________ 10. I have three pairs of legs and three body sections. What kind of arthropod am I?

______________ 11. I am an arthropod. I have five pairs of legs and two pairs of antennae. I also have two body sections. What kind of arthropod am I?

______________ 12. I am an arthropod with two body sections and four pairs of legs. Who am I?

______________ 13. I am an adult insect. I have had to go through three other stages of development before I reached this stage. Who am I?

______________ 14. I am an arthropod with many body segments. On many of my body segments I have a pair of legs. Who am I?

______________ 15. I am an arthropod with many body segments. I have two pairs of legs on each segment. Who am I?

Name ______________________________ Date ______________

DEVELOPMENT OF A FROG

Look at the pictures. Each one shows a different stage in the growth of a frog, from egg to adult. Read about each stage. Then match each stage with the right picture. Write the correct number on the line under each picture.

1. The egg, covered by a jellylike substance, is fertilized.
2. The egg cell divides to make two cells.
3. The cells divide again and again. A ball of cells forms.
4. The embryo changes shape. The head and tail form. The egg hatches. Gills develop and are used for breathing. The embryo is now a tadpole.
5. The tadpole grows bigger. A flap of skin grows over the gills. The hind legs develop.
6. The front legs appear. The tail gets shorter. Lungs form for breathing. The gills disappear.
7. The tail disappears. The tadpole becomes a frog that can live on land.

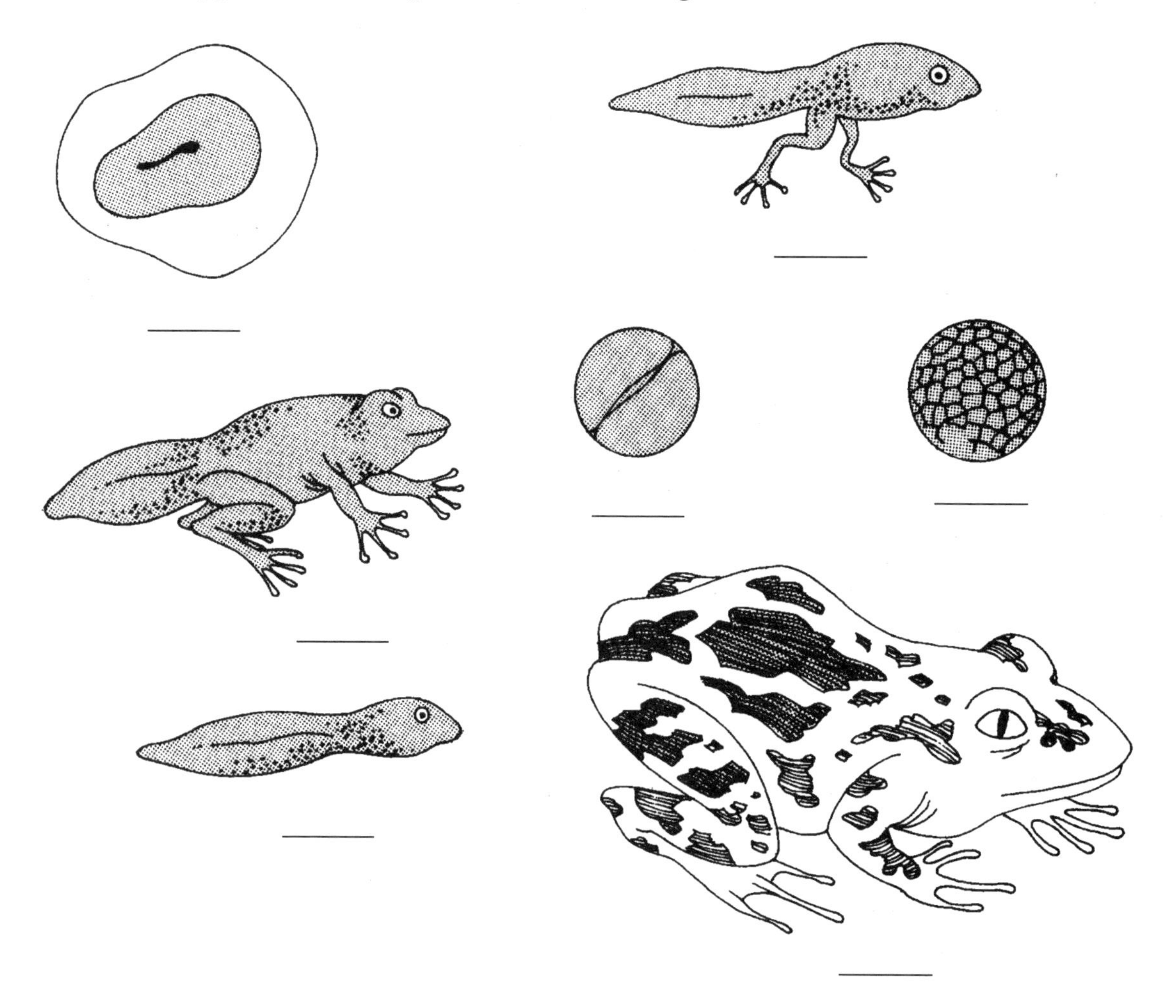

Name ______________________ Date ____________

Fish and Amphibians

You have learned that animals and plants go through a life cycle. A cycle is a period of time in which plants and animals are born, live, reproduce, and die. The word *cycle* comes from the word that meant *wheel* to the ancient Romans and Greeks. A cycle in nature is like the rim of the wheel. If you start at one point of the rim and go around the wheel, you come back to the point from which you started.

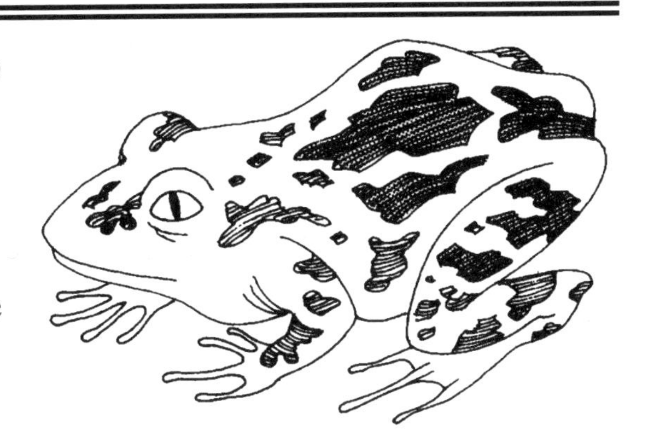

The frog is an amphibian that has an interesting life cycle.

Below is a list of the events in a frog's life. They are not in the correct order. Rewrite the list in the correct order. Then, on another sheet of paper, make a drawing that shows these events.

The tadpoles with gills and tails live in water.
An adult frog lays eggs once again.
The animals leave the water and live as frogs.
Tadpoles lose their gills and tails and develop lungs and legs.
Tadpoles hatch from eggs.

1. ____

____

2. ____

____

3. ____

____

4. ____

____

5. ____

____

Name ______________________________ Date ______________

You Can Raise and Observe Tadpoles

Have you ever searched the edge of a local stream or pond for frog eggs? You may have found frog eggs as clumps of jellylike material. Toad eggs are like strands of jellylike material. In this activity, you will raise tadpoles.

Materials:

- **plastic bucket to scoop up frog or toad eggs**
- **two large rocks**
- **two jars with tops in which you have made holes for air circulation**

Do This:

1. With your teacher, find frog or toad eggs at a pond.
2. Place the eggs, algae from the pond, and pond water in the bucket.
3. In the classroom or at home, remove some of the eggs from the jar and place a few in each of two smaller jars containing pond water and algae. To each add a large rock that extends above the surface of the water.
4. Place one jar in a cool place, and leave one at normal room temperature.

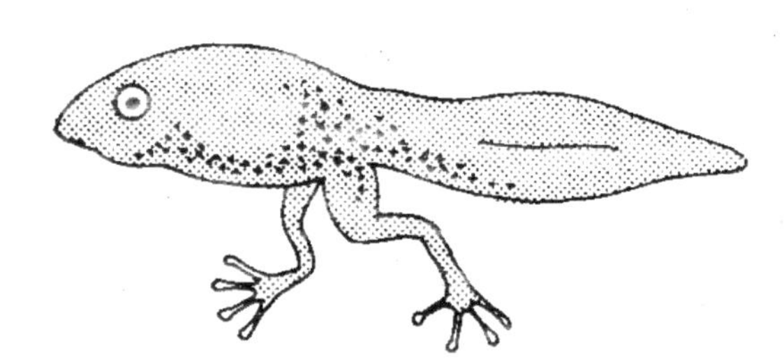

Answer these questions.

1. Make a chart to record your observations of what happens in each container. Make observations once every day for several days.

2. Did any eggs hatch? ______________________________

3. Is there a difference in the two containers? ______________________________

4. The eggs you found had dark and light spots. Based on your observations, what were the spots?

____

____

____

Return all the tadpoles and the unhatched eggs to the pond or stream in which you found the eggs when you are finished.

Name ______________________ Date ______________

OBSERVING FISH IN AN AQUARIUM

Materials:
- aquarium with several fish
- one sheet of paper
- pencil

Do This:

1. Observe the fish as they swim through the water. Notice how their tails and fins move. How do the fish push themselves through the water?

   ____

   How do the fish turn? ______________________________

   How do they come to a stop? ______________________________

2. With a pencil, gently tap one side of the aquarium below water level.

   How do the fish react? ______________________________

3. Wait a few minutes after step 2. Hold the sheet of paper against one side of the aquarium. Tap gently on the paper with a pencil.

   How do the fish react? ______________________________

Answer this question.

Write a paragraph explaining which senses you think the fish were using in steps 2 and 3.

____

____

____

____

____

____

____

Name ______________________ Date ____________

Reptiles and Birds

Complete these exercises.

A. How would you like to have this creature for a pet? Sorry, this animal isn't around anymore. Its name was *Tyrannosaurus. Tyrannosaurus* was 15 m (about 50 ft) long and 5 $^1/_2$ m (about 18 ft) tall. Scientists have found many fossil remains of *Tyrannosaurus*. They have concluded that *Tyrannosaurus* was a reptile. Describe the characteristics that this animal has that led scientists to reach this conclusion.

1. Was it warm-blooded or cold-blooded? ______________________
2. How did it reproduce? ______________________
3. What kind of skin did it have? ______________________
4. *Tyrannosaurus* could live entirely on land. What characteristics that you named above helped it do this?

   ____

B. Below are drawings of two interesting animals. One is an extinct flying reptile called *Pteranodon*. It probably lived about 200 million years ago. A portion of its skin stretched across bone to form a wing. It also had hollow bones that made it lighter. The other drawing shows a modern bird that is in danger of becoming extinct. It is a California condor. Compare the characteristics of the two. Write each animal's characteristics on the lines at the right.

condor

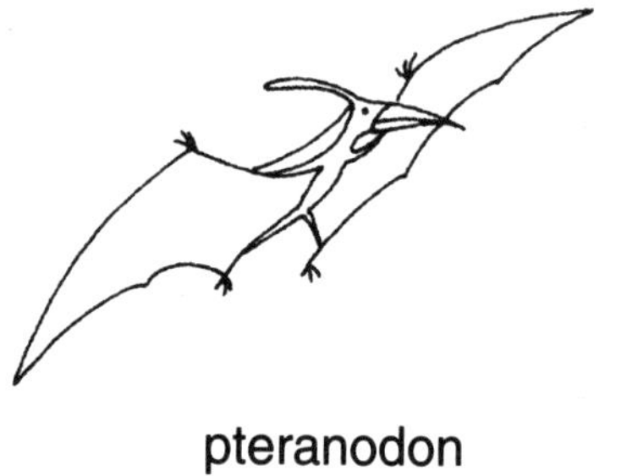

pteranodon

Name ______________________________ Date ______________

Birds

Read the following story about birds. Then follow the directions.

Birds are the only animals that have feathers. This body covering is actually a special kind of scale. Feathers provide insulation. They keep birds from losing too much body heat on cold days.

Because they have hollow bones, birds are very lightweight. This makes it easier for them to fly. They also have a very large breastbone in their skeleton. The powerful muscles that move the wings are attached to it.

Their brains, called eye brains, have well-developed areas that deal with sight. Even when they are high in the sky, birds can see things on the ground quite clearly. However, birds do not have a good sense of smell. This is because the areas of their eye brains connected with the sense of smell are poorly developed.

When in flight, the body of a bird uses a lot of energy and gives off a lot of heat. To cool themselves, these animals have an internal "air-conditioning system." It is made up of a network of air sacs that are connected to the lungs. When a bird's body temperature gets too high, cooling air flows through its air sacs.

Match each sentence with the correct body part. Write the letter of the body part on the line before the sentence.

______ **1.** These cover the body of a bird.

______ **2.** These kinds of brains are well-developed in areas to do with sight.

______ **3.** These help keep birds very light in weight.

______ **4.** Powerful wing muscles are attached to this large bone.

______ **5.** These help cool a bird during flight.

a. breastbone
b. air sacs
c. feathers
d. hollow bones
e. eye brains

Name ______________________ Date ______________

MAMMALS

Complete these exercises.

A. The drawings show two mammals. Look at the pictures and then answer the questions below.

1. Compare the location of the eyes for each mammal.

____

____

2. Which mammal has the best ability to chew through objects? ______________

3. Which mammal has the best ability to hold objects? ______________

4. With what group of animals should the mouse be classified?

____

5. With what group of mammals should the chimpanzee be classified? ______________

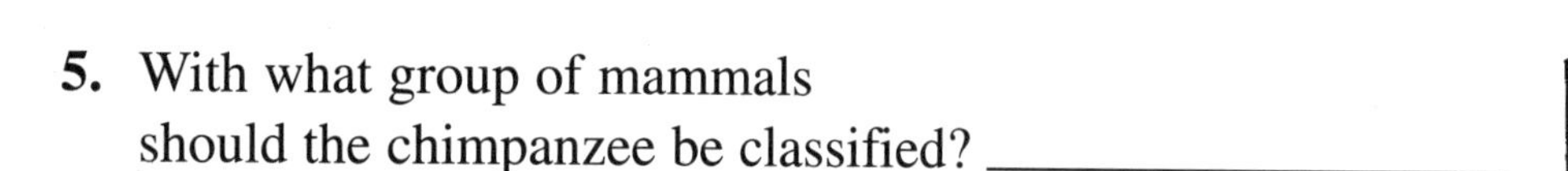

B. All mammals live and grow in their mother's body before they are born. This is called the *gestation period*. The amount of time they develop in the mother's body is different for different mammals. Use these data to make a graph that shows these differences. Use another sheet of paper to make your graph.

Opossum	12 days	Pig	10 weeks
Hamster	16 days	Cow	9 months
Mouse	21 days	Chimpanzee	9 months
Rabbit	30 days	Human	9 months
Cat	9 weeks	Horse	11 months
Dog	9 weeks	Elephant	18-21 months

Study your graph. Then make some predictions about the length of time some other animals remain in their mother's body. Do library research to compare your predictions with the actual lengths of time.

Name ______________________ Date ____________

AN ELEPHANT STORY

Animals adapt to their environments over long periods of time. Read how the elephant developed different traits. The elephants with the traits best fitted to their environments survived, while the elephants without these traits did not. The animals that survived passed on these traits to their offspring. This is called *natural selection.*

Read the following story about elephants. Then follow the directions.

Long ago, many different kinds of elephants lived on the Earth. The earliest elephants were only about the size of a pig. They looked more like modern-day hippopotamuses than like elephants. They had long, heavy bodies, wide feet, short legs, and tiny ears. They also had very long heads. One set of upper teeth was enlarged into small tusks. Instead of trunks, these animals had long noses.

As time passed, several different types of elephants appeared. They were larger than the first elephants. Their skulls, ears, and tusks were larger. Their upper lips and noses had become a short, well-developed trunk. Their legs had also lengthened. These animals looked less like hippos and more like elephants today.

All the early and intermediate types of elephants have become extinct. Today, only two kinds of elephants exist—the African and Asian. They both have long legs and long curving tusks. Their bodies are bigger than the earlier elephants. Their ears are very large. Their trunks are long. Almost no hair grows on their leathery skin, although a tuft of hair is found at the end of their tails.

Decide which of the three kinds of elephants each sentence describes. Write 1 for the early elephant, 2 for the intermediate elephant, and 3 for the modern elephant.

______ It was about the same size as a pig.

______ This elephant had short legs.

______ Its tusks are large and curving.

______ A short, well-developed trunk extended from its face.

______ It had small tusklike teeth.

______ This elephant had a large nose but no trunk.

Name ______________________ Date ____________

Observing Small Mammals

If you are able to care for gerbils or hamsters in school or at home, you may wish to do this activity. If you are not able to care for these animals, you can observe hamsters or gerbils in a pet store.

Materials:

- **cage**
- **water**
- **food feeder**
- **hamster or gerbil food**
- **cage bottom material such as wood chips**
- **two gerbils or two hamsters (male or female)**
- **a book on care and feeding of gerbils or hamsters**

Do This:

1. Set up the cage with food and water. Place the animals in the cage.
2. Observe the animals each day for a period of two weeks.
3. Keep a chart for all of your observations.

Life Science 5, SV 3845-X

Name ______________________ Date ____________

DESCRIBING ANIMALS

Read the list of animals and their descriptions. Then draw a line from each animal to its description.

1. Bird	**a.** amphibian
2. Bat	**b.** sea-living mammal
3. Squirrel	**c.** warm-blooded and lays eggs
4. Frog	**d.** flying mammal
5. Chimpanzee	**e.** fish with skeleton made of cartilage
6. Lizard	**f.** pouched mammal
7. Salmon	**g.** rodent
8. Seal	**h.** primate
9. Opossum	**i.** bony fish
10. Shark	**j.** cold-blooded, lays eggs on land

Name ______________________ Date ____________

CLASSIFYING ANIMALS

The drawing below shows many of the animals you have learned about. Look at the drawing. Then classify each animal by listing its name under the proper heading.

Fish
Amphibians
Reptiles
Birds
Mammals

Name ______________________________ Date ______________

ANIMAL GROUPS

Circle the animal that does not belong with the rest. Then, explain why in the blank.

1.	platypus	turtle	bird	cow
2.	bird	crocodile	frog	trout
3.	snake	shark	horse	ostrich
4.	penguin	bat	eagle	dog
5.	ray	frog	worm	deer
6.	shark	salamander	alligator	man
7.	barracuda	tiger	boa constrictor	elephant

8.

Name ______________________________ Date ______________

TEST YOUR ANIMAL KNOWLEDGE

These exercises will test your animal knowledge. You may use a science book or an encyclopedia to help you find the answers.

A. Arrange the items below in the proper order.

1. By increasing complexity: human, kangaroo, trout, frog, snake

____

2. By number of chambers in heart: salamander, ray, cat

____

3. By ability to live in different environments: human, tuna, frog, lizard

____

B. Identify the types of animals described.

4. I have hands with movable fingers with flat nails. I have hair and two eyes that look forward. I am a ______________________________.

5. I lay eggs, and I cannot control by body temperature. I have a three-chambered heart. My skin is hard and scaly. I am a ______________________________.

6. I lay eggs and cannot control my body temperature. I have a two-chambered heart, and my skin is usually covered with scales. I breathe through gills. I am a ______________________________.

C. In the space below, draw a picture of one of the organisms described in B.

Name ______________________________ Date ______________

TRILOBITES

Millions of years ago, the land was bare and rocky. Few plants and animals lived on the land. The oceans, however, teemed with life. Then, trilobites were common organisms. Today these animals are extinct.

The body of the trilobite was divided into three parts: head, body, and tail. Two antennae that served as sense organs projected from the head. The body was divided into a middle lobe with two side sections.

Trilobites had hard shells that protected them. Fossils show that some trilobites could curl up into balls to protect themselves. You can make a model trilobite.

Materials:

- clay
- table knife
- wax paper
- pencil
- rolling pin

Do This:

1. Divide the clay into two equal parts. Place one part on a piece of wax paper. Roll it flat using the rolling pin. With your knife, cut out an oval piece for the body of the trilobite. Save the leftover scraps of clay.

2. Take the second part of the clay. Roll it into a thick snake the same length as your oval. Shape it so that one end is thinner than the other. Place the snake on the center of the oval. Be sure the thicker end is toward the top or head end of the trilobite. Press the snake onto the oval. With a pencil, scratch out rows of scales along the shell.

3. With the clay left over from step 1, roll out a long, thin snake. Cut it into three pieces. To make the antennae, use two of the cut pieces and press them onto the oval body. To make the tail, cut the remaining part in half. Place the two small pieces close together on the oval.

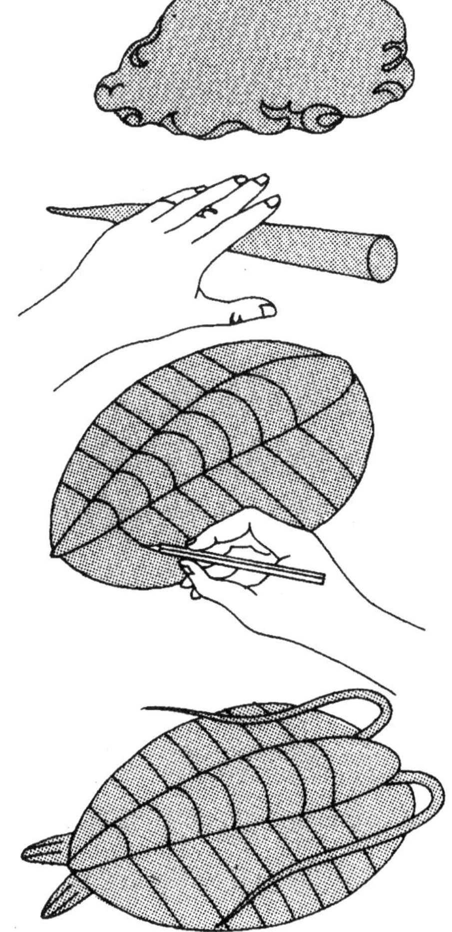

Trilobites were once abundant sea organisms. Today they are extinct. We know about them only from fossils found in sedimentary rocks.

Name ______________________ Date ____________

Ancient Nebraska

Read this story about an event of ancient times. Then answer the questions.

The time was ten million years ago. The place was the flat grasslands of what is now northeastern Nebraska.

Mother rhinoceroses and their babies cool themselves in a water hole. A few grown males wade nearby. Camels, three-toed horses, and saber-tooth deer graze along the banks of the water hole. Wading birds drink from the pool. Turtles swim underwater, among tiny water plants.

In the distance a great black cloud billows toward the unsuspecting animals. Gritty ash begins to fall from the sky. The animals become restless. The ash burns their eyes. It sticks to their nostrils, tongues, and throats. The animals cough and choke.

As the cloud gets closer, the air becomes thick with ash. The smaller animals begin to suffocate. More and more ash falls. The rhinos start to fall, too. Eventually all the animals around the water hole die. For almost a month, the ash rains down on the animals. Now buried beneath a blanket of ash, their skeletons will be preserved for ten million years.

The ash cloud came from a gigantic volcanic eruption far from the water hole. Winds blew the gritty ash away from the volcano. It settled on what is now northeastern Nebraska.

At the site of the ancient water hole in Nebraska, scientists dug up more than 200 brittle skeletons. To protect the skeletons, the scientists made a plaster copy of each bone. They recorded exactly where every animal fell. Scientists used the information to figure out just what happened at the water hole so long ago.

1. When did the rhinoceroses live in Nebraska? ______________________

2. What other animals lived in and near the ancient water hole? ______________________

   ____

3. What caused the ash cloud? ______________________

4. What did the ash do to the animals? ______________________

   ____

5. How did the scientists protect the bones that they dug up?

   ____

Name ______________________ Date ______________

Fossil Clues

Each paragraph below gives facts about different fossils. The conclusions at the bottom of the page tell what scientists have learned from these facts. Match each paragraph with the conclusion that scientists reached. Write the letter of the conclusion after the paragraph.

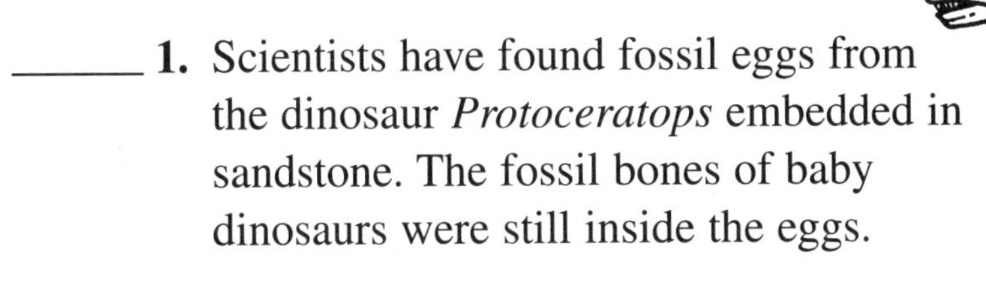

______ **1.** Scientists have found fossil eggs from the dinosaur *Protoceratops* embedded in sandstone. The fossil bones of baby dinosaurs were still inside the eggs.

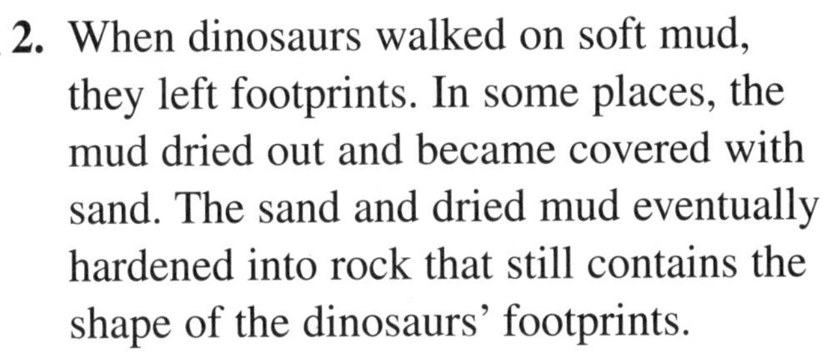

______ **2.** When dinosaurs walked on soft mud, they left footprints. In some places, the mud dried out and became covered with sand. The sand and dried mud eventually hardened into rock that still contains the shape of the dinosaurs' footprints.

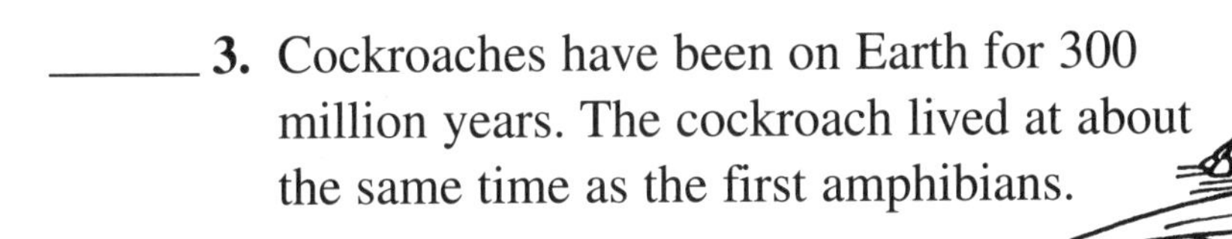

______ **3.** Cockroaches have been on Earth for 300 million years. The cockroach lived at about the same time as the first amphibians.

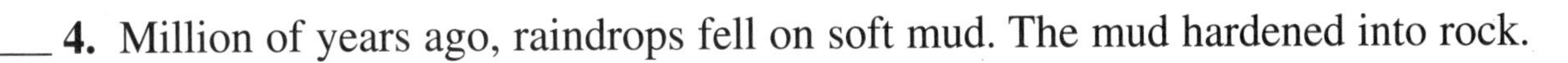

______ **4.** Million of years ago, raindrops fell on soft mud. The mud hardened into rock.

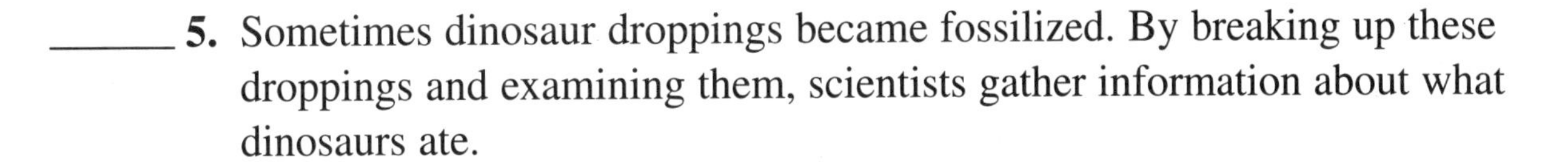

______ **5.** Sometimes dinosaur droppings became fossilized. By breaking up these droppings and examining them, scientists gather information about what dinosaurs ate.

Conclusions:

A. Some dinosaurs ate plants. *Brachiosaurus*, a giant plant-eating dinosaur, must have eaten about a ton of leaves every day. Other dinosaurs ate animals. *Tyrannosaurus* was the largest of the meat-eating dinosaurs.

B. Some dinosaurs buried their eggs in sand.

C. Tiny circles in rocks give clues about the climate. They show where rain was followed by a dry period.

D. *Megalosaurus*, a huge dinosaur, had three toes.

E. Amphibians of 300 million years ago may have eaten insects such as cockroaches.

Name ______________________ Date ____________

THE WEB OF LIFE

Read the information below. Watch for the words in the box.

biosphere	**populations**	**communities**	**food chains**	**balance**
predators	**prey**	**producers**	**consumers**	**decomposers**

The biosphere includes the atmosphere, the Earth's crust, the oceans, and all the life that exists there. The Sun is also considered part of the biosphere. Its energy enters the atmosphere and is important to life in the biosphere.

Within the biosphere are many different populations of animals and plants. Populations interact with other populations and nonliving things in their environments. Groups of populations living together form communities. The populations within these communities form food chains. These populations depend upon each other for their survival. The food chains create a balance in the community.

Here is an example. In a forest community, there are different populations. There are grasses, rabbits, and foxes. The rabbits eat the grasses. The foxes eat the rabbits. If there were no foxes, the rabbit population could grow too quickly. They would use up all the grasses. Then the rabbits would begin to die because there would not be enough grasses. If there were no rabbits in the community, the foxes would die. They could not live on the grasses in the woods. Because there are predators and prey in the community, the community stays balanced.

Every community has three types of organisms. There are producers, consumers, and decomposers. The producers are usually plants. The consumers are usually the animals. Decomposers are molds and bacteria. They break down the dead organisms and return them to the soil. All of these keep the community balanced. If something hurts a population, like a disease, a fire, or pollution, it affects the whole community.

Go on to the next page.

Name ______________________ Date ______________

THE WEB OF LIFE, P. 2

Complete these exercises.

1. Make a diagram showing a decomposition>producer>consumer cycle. Use the following labels in your diagram.
   - **a.** animals and plants die
   - **b.** decomposers break down dead organisms
   - **c.** materials return to the soil and are used by plants to make new food
   - **d.** animals eat plants
   - **e.** animals eat animals

2. Draw a picture of a predator and its prey.

Name ______________________ Date ______________

DIFFERENT KINDS OF CONSUMERS

Consumers are divided into three categories. Consumers that eat only plants are called *herbivores*. Those that eat only animals are called *carnivores*. Those that eat both plants and animals are called *omnivores*.

Label each picture below as *herbivore, carnivore,* or *omnivore*. You may use a science book or encyclopedia if you are not sure.

____

____

____

____

____

____

____

____

____

____

____

____

Name ______________________________ Date ______________

DECOMPOSERS

Along with plants and animals, decomposers are important to the balance of nature. The decomposers, such as mold and bacteria, break down dead organisms and waste and return them to the soil. Plants then use the soil to grow. This completes the cycle. Where can you find decomposers? What do they need in order to be active? To find out, try this activity.

Materials:

- **four plastic containers**
- **marking pen**
- **newspaper**
- **potting soil**
- **sand**
- **water**

Do This:

1. Label the four containers *dry soil*, *damp soil*, *dry sand*, and *damp sand.*
2. Tear four strips of newspaper about 4 cm wide and 12 cm long. Lay each strip in a plastic container, with one end of the strip hanging over the edge.
3. Fill the two containers labeled *dry soil* and *damp soil* with potting soil. Fill the other two containers with clean sand. One end of the strip of newspaper should stick out of the soil or sand in each container.
4. Add a little water to the containers labeled *damp soil* and *damp sand.* Add water to the containers every few days to keep the soil and sand damp.
5. Observe the condition of the strips of newspaper in the containers for one week. Record your observations below.

Day	Dry Soil	Damp Soil	Dry Sand	Damp Sand
1				
2				
3				
4				
5				
6				
7				

Go on to the next page.

Name ______________________ Date ______________

DECOMPOSERS, P. 2

Answer these questions.

1. In what container did the strip of newspaper decompose the most? Explain why you think this happened.

____

____

____

2. What happened to the strip of newspaper in the containers that were kept dry? Explain why you think this happened.

____

____

____

3. What could account for the difference in decomposition in the container with the damp soil and the one with the damp sand?

____

____

____

4. How do decomposers help the environment?

____

____

5. Add arrows to the diagram below to show how energy flows through a food web.

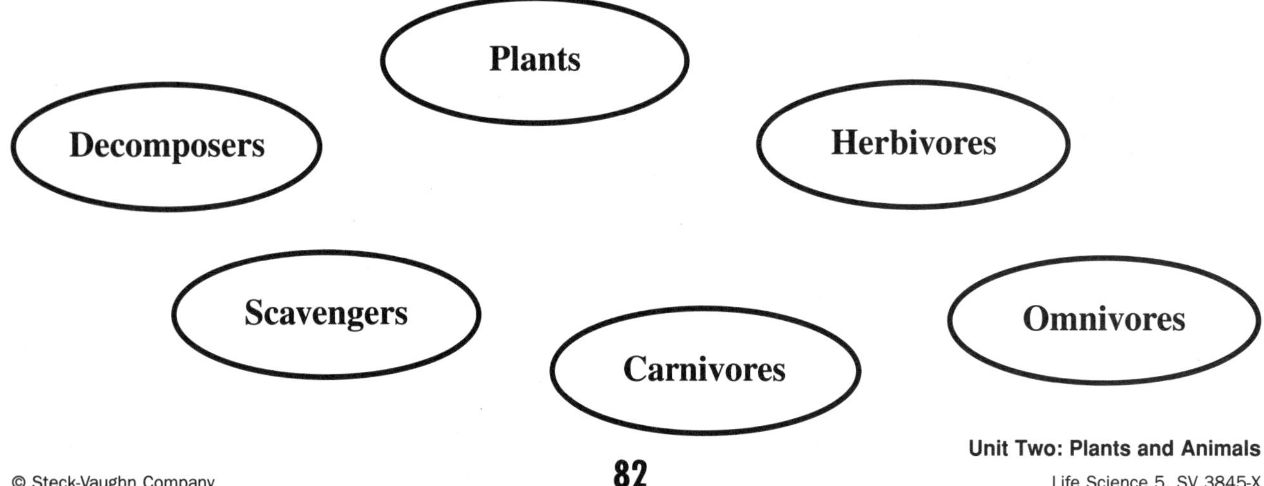

Name ______________________ Date ____________

Plants and Animals Adapt

Plants and animals change over time to become fitted to their environments. These changes are gradual. Plants and animals cannot adapt to meet sudden changes. Sudden changes, such as erupting volcanoes or poisons in the environment, can harm plants and animals. They are not able to change in ways to avoid these dangers. However, plants and animals do change over thousands of years.

One way that plants and animals can adapt to their environment is to move. When a population of animals moves, it is called *migration.* When a population of plants move, it is called *succession.* Animals usually move according to the seasons. Many birds fly south in the winter to keep warm. Then they fly north in the spring to reproduce. When plants move, it is usually to repopulate an area. If a volcano erupts on an island, for example, and kills all the vegetation, plant life will eventually appear again. The plants move onto the island.

Animal adaptations can be seen easily. Many animals are colored so that they will blend into their environments. This makes it difficult for predators to spot them. Others are brightly colored so that they can attract a mate. This ensures that they will reproduce.

Birds have beaks that are adapted to the type of food they eat. Some have short, strong beaks. Others have long, scissorlike beaks. Other animals have teeth that are fitted to the foods they eat. Meat-eating animals have sharp teeth for tearing meat. Animals that eat grasses have flat teeth for grinding.

Many more adaptations help animals survive. Claws are important to some animals for catching their food. The long neck of the giraffe allows it to eat the leaves from high up in the trees. The strength and speed of the lion help it to catch and overpower its prey. In turn, the speed of the zebra and the gazelle can help them to get away.

From the tiniest organisms to the largest, all are adapted to their environments.

Go on to the next page.

Name ______________________________ Date ______________

Plant and Animals Adapt, p. 2

Complete these exercises.

Choose a plant or animal that interests you. Research this organism to see where it lives, what it eats, and how it reproduces. Write about what you learn. Write about the characteristics of the organism that help it to live. What does it look like? How does it eat? How does it move? Try to find out how long it has been Earth. How has it changed over the years? Is it endangered? Why? Draw a picture to go with your report.

____

____

____

____

____

____

____

____

____

____

____

____

____

____

____

____

____

Name ______________________ Date ____________

COMPARING BIOMES

A biome is a large community of plants and animals. Biomes are characterized by the plants and animals found there and by a specific type of climate. Different parts of the world with the same type of biome may have very similar climates. For example, most deserts are extremely dry and have very little rainfall, no matter where they are located.

Look at the following bar graph to find the amount of precipitation each biome receives annually.

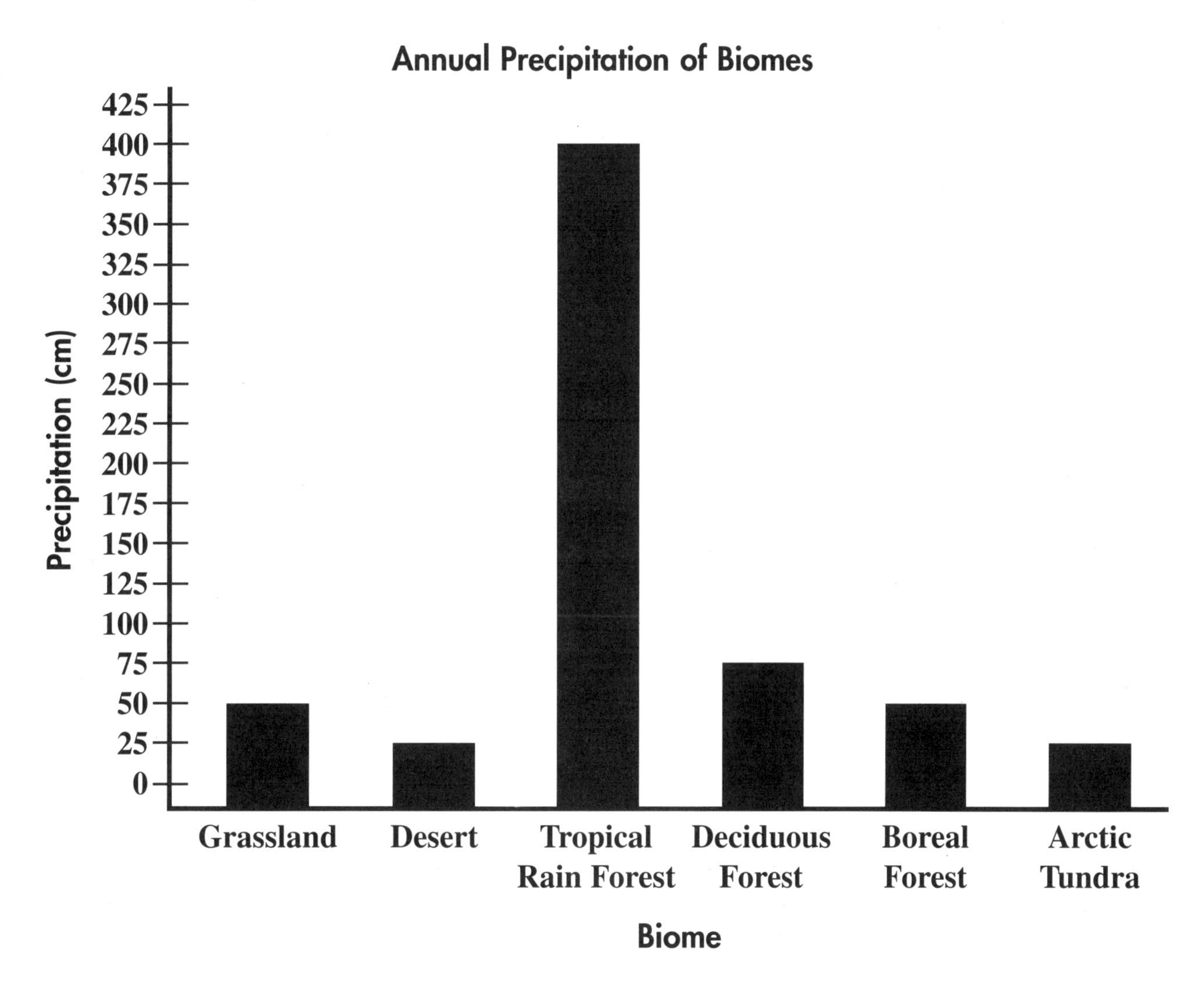

Go on to the next page.

Name ______________________________ Date ______________

Comparing Biomes, p.2

Answer these questions.

1. Write the amount of precipitation next to the name of each biome.

Grassland ____________________

Desert ____________________

Tropical rain forest ____________________

Deciduous forest ____________________

Boreal forest ____________________

Arctic tundra ____________________

2. Which biome has the greatest amount of rainfall in a year? the least amount of rainfall?

____

____

3. Which biome—deciduous forest or boreal forest—receives more rainfall?

____

4. Judging from the amount of rainfall in a tropical rain forest, what do you think the climate would be like there?

____

____

____

5. Deserts receive very little rainfall. What kinds of plants and animals do you think could live in a desert biome?

____

____

____

Name ______________________ Date ______________

Earth's Biomes

Read the paragraphs and then answer the questions that follow.

A biome is a large community of plants and animals. The type of biome is determined by climate and the kinds of plants found there. There are six major land biomes in the world.

Tropical rain forests grow where the climate is warm and rainy. Most of the animal life is found in the trees. Rain forests are important because they produce oxygen, which supports life.

Deciduous forests have warm or hot summers and cold winters. Deciduous trees lose their leaves every fall. Deer, squirrel, foxes, owls, and snakes are found in deciduous forests.

Boreal forests grow in places with very cold, snowy winters and short growing seasons. The trees are mostly evergreen. In this type of forest you will find deer, bears, snowshoe hares, and beavers.

On the **Arctic tundra**, the winters are long and cold, and the summers are short and cold. Animals on the tundra, such as the snowy owl, are adapted for cold weather. They also blend in with the snow.

Grasslands have winters that are cold and snowy and summers that are hot and dry. Many small animals, such as ground squirrels, prairie dogs, and many kinds of birds, are found in the grasslands.

Because **deserts** receive very little rainfall, the plants there are far apart so they don't compete with each other for moisture. The desert supports many animals, such as mice, snakes, and coyotes.

Go on to the next page.

Name ______________________ Date ____________

Earth's Biomes, p. 2

1. Use the boldface words from the paragraphs you have read to complete the table.

Biome	Description
	long, cold winters; animals blend with snow
	warm summers, cold winters; trees lose leaves
	receive little rainfall; plants are far apart
	cold, snowy winters; evergreen trees
	hot, dry summers; home to prairie dogs
	warm and rainy; produce much oxygen

2. Identify the biome each animal lives in.

______________ ______________ ______________

______________ ______________ ______________

Name ______________________________ Date ______________

People and the Environment

Read the paragraphs and then answer the questions that follow.

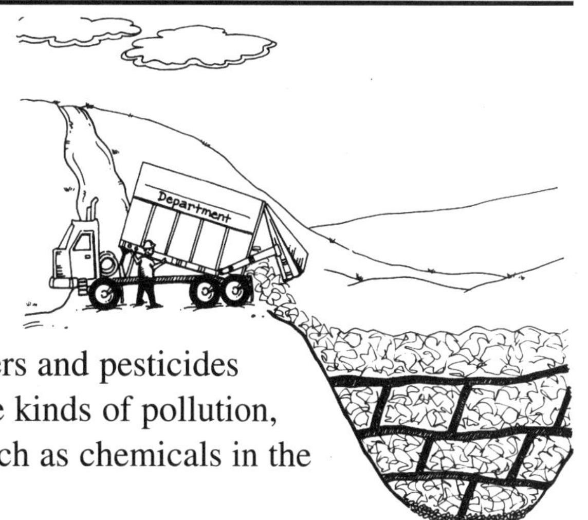

People change the land when they use it. Farming, grazing, and building all cause change. As people use land, they also produce wastes. Wastes can be harmless, or they can destroy the environment. Some farmers use chemical fertilizers and pesticides that can harm the soil and the water supply. Some kinds of pollution, such as landfills, are very visible. Other kinds, such as chemicals in the air, are harder to see.

One way for people to help control pollution is to recycle. Aluminum cans, plastic, and newspaper are just a few things that can be recycled. Another way is for farmers to use alternatives to pesticides, such as natural predators, to get rid of pests.

1. Name four ways that people pollute the land.

______________________ ______________________

______________________ ______________________

2. Name three ways in which people can control pollution.

a. ______________________________________

b. ______________________________________

c. ______________________________________

Go on to the next page.

Name ______________________ Date ______________

PEOPLE AND THE ENVIRONMENT, P. 2

3. Look at the illustration. List all the ways that people have changed the land.

______________________ ______________________

______________________ ______________________

______________________ ______________________

______________________ ______________________

______________________ ______________________

______________________ ______________________

4. Why is it important to recycle?

____

____

____

____

____

Name ______________________ Date ______________

POLLUTION

When we hear the word *pollution*, most of us think of either water pollution or air pollution. But there are many other types of pollution. The most visible type is solid-waste pollution. This includes anything from trash along the highway to junked cars and old tires. Anything solid that you throw away—such as an old comic book or the leftovers from your lunch—is solid waste. And any waste that harms the environment is a form of pollution.

The United States produces about two billion tons of solid waste each year. That's about 20 kg of solid waste for each person each day! Think about how much waste that is—about half your body mass.

What can we do about solid waste? The good news is that with a little effort, we can cut down on pollution. We can recycle.

Think about an aluminum can. You can throw it away, or you can recycle it. When you throw it away, it will probably be taken with the rest of your garbage to a landfill. The smell of a landfill pollutes the air. In some cases, materials in the landfill leak down through the soil and pollute underground water.

Think a little more about the aluminum can. If you recycle it, you won't be adding it to an overfilled landfill. But just as important as that, reusing aluminum means saving natural resources and saving energy. Any aluminum that is recycled will not need to be mined from the ground. Energy that would be needed to process new aluminum ore will be saved.

In the last few years, many communities have started recycling aluminum cans, glass bottles and jars, and newspapers. Some communities have also begun to recycle other waste, such as tin-covered steel cans and plastic containers. Each item that is recycled saves energy and doesn't add to solid-waste pollution. You can help. Find out how much you and your family can recycle.

Materials:

- **paper grocery bags**
- **marking pen**
- **trash**

Do This:

1. Keep track of all the things you can recycle in one week. Before you throw something away, think about whether it could be recycled.

Go on to the next page.

Name ______________________________ Date ______________

POLLUTION, P. 2

2. Organize your recycling. Label a paper grocery bag for each type of item that can be recycled. You should have one bag for each of the following: newspapers; magazines and junk mail; cardboard; glass; aluminum; and tin.

3. Ask your family to help by putting anything that can be recycled into the right bag.

4. At the end of the week, fill out the following table to find out how much you recycled. Count the number of cans and bottles in each category. For the paper and cardboard recycling, you can weigh each type if you have a scale. Otherwise, count the number of newspapers, magazines, and cardboard boxes.

5. Most communities recycle some things. Find out whether your community will accept the items you separated for recycling.

Newspaper	Cardboard	Magazines and Junk Mail	Glass	Aluminum	Tin

Answer these questions.

1. If recycling is so helpful, why do you think many people do not recycle?

____

____

____

2. What can you do to help prevent solid-waste pollution?

____

____

____

Name ______________________ Date ____________

UNIT 2 SCIENCE FAIR IDEAS

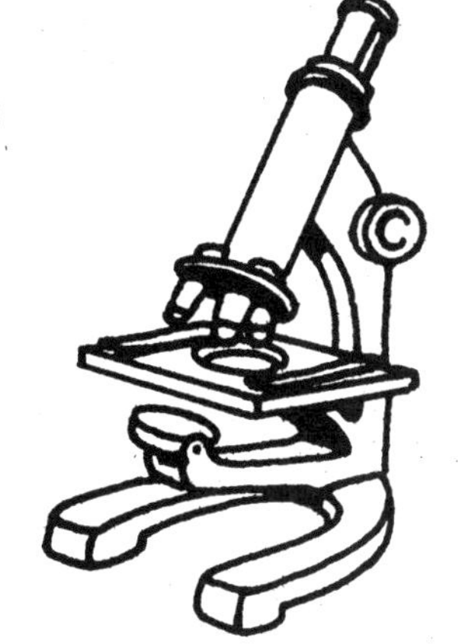

A science fair project can help you to understand the world around you better. Choose a topic that interests you. Then use the scientific method to develop your project.
Here is an example:

1. **PROBLEM**: How are animals fitted to their environment?

2. **HYPOTHESIS**: Animals have adapted to their environments over the years.

3. **EXPERIMENTATION**: Examine several different animals. See how their color, size, shape, or habits help them to survive in their environment.

4. **OBSERVATION**: Animals have developed special ways of surviving in their environments. Some animals use camouflage for safety. Some migrate to warmer climates in the winter. Some are tall to reach foods high in trees. Some are fast to catch prey or to escape from predators. Animals have beaks, teeth, and/or claws that are fitted to the type of food they eat. Animals are fitted to their environments in many ways.

5. **CONCLUSION**: Animals have adapted to their environments.

6. **COMPARISON**: Conclusion agrees with hypothesis.

7. **PRESENTATION**: Display pictures and facts about the animals you researched. Tell what special traits of each animal help it to survive in its environment.

8. **RESOURCES**: Tell of any reading you did to help you with your experiment. Tell who helped you to get materials or set up your experiment.

Other Project Ideas

1. How do plants and animals help each other to survive?
2. What happens to dead plants and animals?
3. How do different plants reproduce?
4. How do plant populations move from one place to another?
5. How do populations become endangered?

Unit 3: Your Body, Inside and Out
Background Information

The Human Body

The cell is the first level of organization in the human body. Groups of cells that have the same structure are called *tissues*. An organ is a group of different kinds of tissues working together to do a job. A system is a group of organs working together to do a job.

The human body has ten systems: **Circulatory System**: transports materials to all parts of the body; **Muscular System**: makes the body move; **Endocrine System**: regulates growth and development, helps control some body functions; **Skeletal System**: gives the body shape and support, protects inner organs; **Integumentary System**: skin, hair, and nails protect the body; **Reproductive System**: enables adults to produce offspring; **Respiratory System**: takes in oxygen and releases carbon dioxide; **Digestive System**: breaks down food into nutrients for cells to use; **Excretory System**: removes waste produced by cells; **Nervous System**: controls the body and helps it respond to the environment.

Each of these systems is included in this unit. Some of them are mentioned briefly, while others are given more space. The amount of information necessary for the teacher to be successful in teaching these pages is complete on the pages themselves and in the answer key. Students are encouraged to use resource books to find answers to questions.

The Five Senses and the Nervous System

The human body collects information using the five senses—sight, smell, hearing, taste, and touch. The nervous system enables us to put all of our senses together so that messages are sent to the brain and we are able to act according to the information that the brain receives. The nervous system enables us to react. It controls all of the other systems in the body.

The major organ of the nervous system is the brain. Another part of the nervous system is a system of nerves that carry information to the brain. The third part of the nervous system is the sense organs. The nose is the sense organ for the sense of smell. There are many nerve cells in the nose that take the information regarding odors to a main nerve called the *olfactory nerve*. The olfactory nerve carries the information to your brain. Your brain will then tell your body what to do with the information.

Muscles, Bones, and Joints

The human body has more than 600 muscles. Muscles enable us to move, keep some organs moving, and connect bones and skin together. Our muscles keep our blood moving, help us to digest, and keep our lungs expanding and contracting.

Some muscle movements are voluntary, and some are involuntary. Most voluntary muscles are connected to bones. (Tendons are tough cords that connect muscles to bones when they are not directly connected.) When you want to move your arm, you move your biceps and triceps. These muscles contract, and your arm moves. A sheet of muscles under your lungs moves in and out without your conscious effort. These muscles make up the diaphragm. The movement of the diaphragm causes air to rush in and out of your lungs. This movement is involuntary. Some muscle movements are not wholly voluntary or involuntary. Can you control the blinking of your eyes? If you try to stop blinking altogether, you will see that you do not have complete control. Your eye will blink.

When you decide you want to move, a message is sent to your brain. Your brain sends a message to the appropriate muscle to contract. The muscle shortens and becomes firm, and the movement occurs. When you want to stop the movement, your brain tells the muscle to relax. People who are physically fit have muscles that are never fully relaxed. They are always slightly flexed and firm. This is called *muscle tone*. To get muscle tone, large amounts of blood need to be supplied to the muscle cells. In order to get the blood to the muscle cells, a person must exercise. Muscles that do not get the necessary blood, or that do not get used enough, become weak and soft.

There are three kinds of muscle cells—smooth, cardiac, and skeletal. The smooth muscle cells are long and thin, and pointed at each end. They have one nucleus. An example of a smooth muscle is a stomach muscle. Cardiac muscles control the heart. The cardiac muscle cells branch out and weave together. They also have one nucleus. The skeletal muscle cells resemble straws and have many nuclei. The tongue and lips contain skeletal muscles.

There are 206 bones in the human skeletal system that support, protect, and move the body. Bones also produce blood cells in their marrow. The marrow is in the hollowed center of the bone. In young people, all bones have red, blood-producing marrow. Older people have red marrow only in the flat bones, such as the ribs. The other bones contain yellow marrow that does not produce blood cells. Cartilage is a soft, rubbery substance that is found where some bones meet. It keeps the bones from rubbing together. You have cartilage at the end of your nose, too.

Bones join in three ways. A hinge joint allows the bone to move back and forth, as the knees and elbows do. A ball-and-socket joint, such as the shoulder joint, allows the body to move in many directions. A ball-shaped bone fits into the hollow of another bone. Pivot joints, like the one that joins the head to the spine, allow the bones to move around and back. Ligaments, strong bands of material that hold the bones in place, join bones at movable joints.

Blood

Adults have about five liters of blood in their body. Children have about four liters of blood. Blood is a tissue that is more than half liquid. The liquid part of blood is called *plasma*. Plasma is mostly water. Red blood cells, white blood cells, and platelets float in the plasma. Red blood cells make up about half of the blood, and the white blood cells and platelets make up the rest. Red blood cells, which resemble tiny flattened balls, carry oxygen from the lungs to the body tissues, and take carbon dioxide from the tissues to the lungs. They are red because they contain a substance called *hemoglobin*. White blood cells, which are larger than red blood cells and irregular in shape, help fight off disease. After the protective covering cells of the skin, hair, and mucous, white blood cells are the body's second line of defense. If bacteria enter the body, white blood cells move toward them and "swallow" them. (The bacteria then leave the body in pus.) Platelets are important in the clotting of blood when the body is injured.

Healthy Bodies

Health for children revolves around healthy foods, plenty of exercise, and good hygiene. As children grow, they should recognize that they can make choices that will help them live healthy lives. They need to learn the connections between what they eat and the way they look and feel. They need to have the basic information that will help them to make good food choices. Children need to know that it is never too early to begin healthy habits in eating, exercise, and hygiene. The habits they form now will affect their lives for many years to come.

Nutrition

The body needs to receive certain nutrients in order to grow and to stay healthy. These nutrients are broken down into six types: carbohydrates, protein, fat, vitamins, minerals, and water.

- Carbohydrates are sugars and starches. Sugars, such as fruits and honey, give the body quick energy while the starches, such as bread, cereal, and rice, give the body stored energy.
- Proteins come from foods such as milk, cheese, lean meat, fish, peas, and beans, and they help the body to repair itself. Proteins are used by the body to build muscle and bone, and they give the body energy.
- Fat is important for energy, too, and it helps to keep the body warm, but if the body does not use the fats put into it, it will store the fat. Fats come from foods such as meat, milk, butter, oil, and nuts.
- Vitamins are important to the body in many ways. Vitamins help the other nutrients in a person's body work together. Lack of certain vitamins can cause serious illnesses. Vitamin A, for example, which can be gotten from foods such as broccoli, carrots, radishes, and liver, helps with eyesight. Vitamin B, from green leafy vegetables, eggs, and milk, helps with growth and energy. Vitamin C, from citrus fruits, cauliflower, strawberries, tomatoes, peppers, and broccoli, prevents sickness.
- Milk, vegetables, liver, seafood, and raisins are some of the foods that provide the minerals necessary for growth. Calcium is a mineral that helps with strong bones, and iron is needed for healthy red blood.
- Water makes up most of the human body and helps to keep our temperature normal. It is recommended that you drink several glasses of water each day.

Foods have long been divided into four basic food groups—meat, milk, vegetable-fruit, and bread-cereal. New discoveries have led to a change in the divisions so that in a food pyramid, fruits and vegetables are separated, and fats are included at the top of the pyramid. The recommended servings for each group have also changed over time. Eating the right amount of foods from each group each day gives one a balanced diet. Eating too many foods from one group or not enough of another can lead to deficiencies or weight problems. Although vitamin supplements can help with these deficiencies, vitamins are best absorbed in the body naturally through the digestion of the foods that contain them.

- The Bread-Cereal (Grain) Group contains foods made from grains such as wheat, corn, rice, oats, and barley. Six to eleven servings from this group each day give you carbohydrates, vitamins, and minerals.
- The Vegetable and Fruit Groups contain vitamins, minerals, and carbohydrates. Two to four servings of fruits and three to five servings of vegetables each day are recommended.
- The Meat Group includes chicken, fish, red meats, peas, nuts, and eggs. The meat group contains much of the protein we get from our diets, but it also includes fats. Two to three servings from the meat group each day are recommended.
- The Milk Group includes milk (whole and skim), butter, cheese, yogurt, and ice cream, and gives us fat, vitamins, protein and minerals that are important for strong bones and teeth, such as Vitamin D. Two to three servings from the milk group each day are recommended.
- The Fats, Oils, and Sweets Group, including butter, oil, and margarine, should be used sparingly.

Hygiene

Keeping the body clean is an important part of staying healthy. Children need to know that when they wash, they are washing off viruses and bacteria, or germs, which can cause illness. Washing the hair and body regularly prevents bacteria from entering the skin through cuts and from getting into the mouth. Hands should always be washed after handling garbage or using the bathroom.

Germs can also come from other people. Children should be discouraged from sharing straws, cups, or other utensils. They should be reminded always to cover their mouths when they sneeze or cough, and to use tissues frequently. Children also need to be reminded not to share combs or hats.

Teeth

Regular brushing and flossing can help keep teeth healthy. Avoiding sweets will also help. Most children have all their baby teeth by the time they are two years old. When they are about seven, they begin to lose their baby teeth and permanent teeth begin to appear. Although the baby teeth fall out, it is important to take good care of them and the gums that surround them.

Decay is caused by acids in the mouth that eat into the enamel. The acids are caused by bacteria that live on the food in your mouth. If you brush and floss regularly, the food is taken out of your mouth, and the bacteria cannot live there. When you brush, you remove the plaque from your teeth, as well. Plaque is the sticky yellow film that develops on your teeth from food, bacteria, and acid. Decay can cause a hole in the tooth called a *cavity*. It can also harm the gums and cause gum disease. Regular dental exams and x-rays will detect any decay that you may have missed.

Name ______________________ Date ____________

We're Organized

As you read the paragraph, look for the words from the box. Then complete the table that follows.

cell	**organ**	**organism**
system	**tissues**	**blood**

The *cell* is the first of five levels of organization of living things. Groups of cells that have the same structure and do the same job are called *tissues*. The muscle in your biceps—in the front of your arm—is an example of a tissue. An *organ* is a group of different kinds of tissues working together to do a specific job. Your heart is an example of an organ. A *system* is a group of organs working together to do a job. Your respiratory system, for example, is responsible for breathing. Your entire body is an example of the highest level of organization—the *organism*.

Classify each thing listed below by using one of these terms: *cell, tissue, organ, system, organism.*

______________	**stomach**	______________	**muscle**
______________	**blood**	______________	**lung**
______________	**leaf**	______________	**oak tree**
______________	**digestive**	______________	**nervous**
______________	**nerve**	______________	**tibia**
______________	**tree bark**	______________	**amoeba**

Name ________________________ Date ______________

THE SYSTEMS

Blood is a tissue composed of several kinds of cells—red blood cells, white blood cells, and platelets. Blood is part of a body system that includes several major organs, such as the heart.

Your body has trillions of cells, hundreds of tissues, dozens of organs, and ten systems: skeletal system, muscular system, digestive system, nervous system, excretory system, respiratory system, circulatory system, endocrine system, reproductive system, and integumentary system.

Use a science book or encyclopedia to label each system in the chart.

System	Function
	This system transports materials to all parts of the body.
	Without this system, you couldn't move from place to place or lift things.
	This system regulates growth and development and helps control some body functions.
	Bones of this system give your body its shape and provide support. They also protect internal organs.
	The skin, hair, and nails of this system provide a protective layer for your body.
	This system provides a way for adults to produce offspring.
	This system takes in oxygen and releases carbon dioxide.
	The food you eat must be broken down by this system into nutrients your body cells can use.
	This system removes the wastes produced by your body cells.
	This system controls your body and helps you respond to your environment.

Name ______________________ Date ____________

SYSTEMS WORKING TOGETHER

Remember:

- Cells are the basic units of all living things.
- Groups of cells with the same function are called tissues.
- Groups of tissues with the same function are called organs.
- Organs with related functions belong to a system.

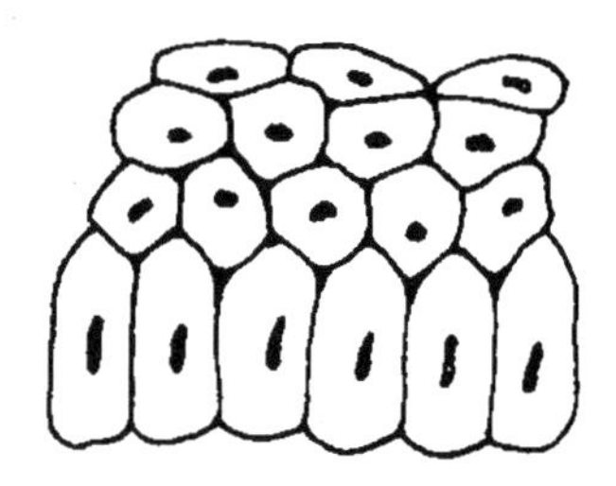

Match the terms at the right with the definitions on the left.

_____ **1.** a body structure made of different kinds of tissues that work together to do a specific job

_____ **2.** groups of cells with the same structure and function

_____ **3.** the basic unit of structure and function of an organism

_____ **4.** fluid tissue that moves from place to place

_____ **5.** a living thing that carries out all life functions

_____ **6.** a group of organs that work together to do a job

a. cell

b. tissues

c. organ

d. system

e. organism

f. blood

Systems in the body work together to get things done. The digestive system, the circulatory system, and the respiratory system work together to provide your body cells with food and oxygen they need to function. The digestive system turns the food you eat into nutrients that the body can use. At the same time, the respiratory system brings oxygen into the lungs. Oxygen passes from the lungs into the circulatory system, and nutrients pass from the small intestine into the circulatory system. In the circulatory system, the blood carries the oxygen and nutrients to all the cells of the body.

7. Can you think of another way that the systems of the body work together?

____

____

____

____

____

Name ______________________ Date ______________

YOUR SENSE OF SMELL

A. Read the following paragraph. Then number the sentences below the paragraph in the correct order.

Arthur's father packed a surprise lunch for him today. Arthur thinks that his sandwich is peanut butter and jelly, as usual. When he sits down at the lunch table and opens his lunch bag, he will immediately know that his dad made him a tuna fish sandwich. How will Arthur know this?

______ The message will be carried to Arthur's brain.

______ Nerve cells in his nose will pick up the message of the odor of tuna fish.

______ The message will be carried to his olfactory nerve.

______ Air containing the odor of tuna fish will enter Arthur's nose.

______ His brain will tell him that the sandwich smells like tuna fish.

B. Rewrite the following science words in order from the simplest to the most complicated:
organ, body system, cell, tissue

____

C. Make a drawing that shows the parts of the body that provide you with a sense of smell. Label each part.

Name ____________________ Date ____________

How Quickly Does Your Nose Detect an Odor?

To find out, try this activity.

Materials:

- **pencil**
- **saucer**
- **record sheet**
- **perfume**
- **clock with second hand**

Do This:

1. Your teacher will divide the class into two groups. Half the class will observe while the other half records data. Then, you will switch roles and repeat the activity.
2. As an observer, you will shut your eyes. Your teacher will then pour some perfume into a saucer.
3. Raise your hand when you first smell the perfume. Keep your hand raised until a hand count has been taken.
4. As a recorder, count and record the number of people who have their hands raised at each 15-second interval. Then enter your data on the chart.
5. Graph the data from your chart.

Seconds	Number of People
15	
30	
45	
60	
75	
90	
105	
120	
135	
150	
165	
180	

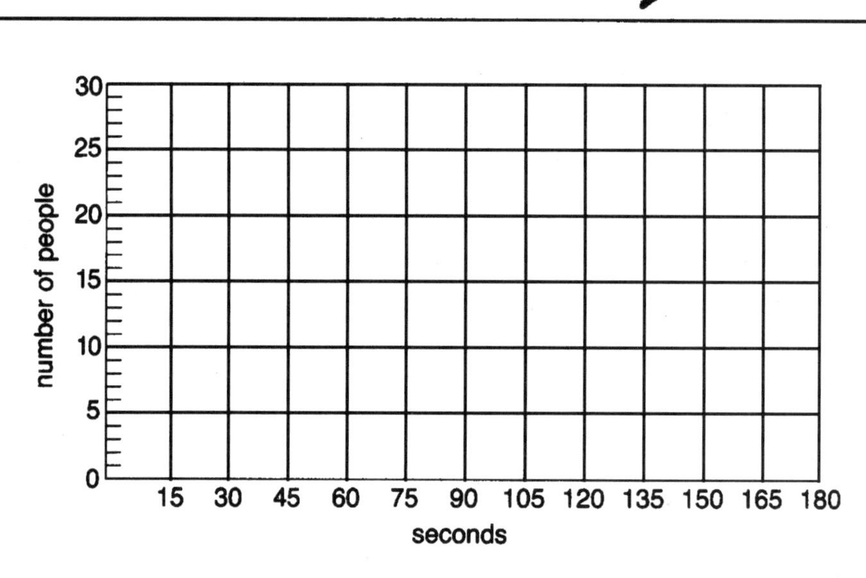

Go on to the next page.

Name ______________________________ Date ______________

How Quickly Does Your Nose Detect an Odor?, p. 2

Answer these questions.

1. How long did it take the first observer to notice the perfume odor?

____

2. How long was it before everyone noticed the odor?

____

3. Why did observers take different amounts of time to detect the odor?

____

____

____

____

____

____

Do the experiment once more with the perfume at the opposite side of the room. Have all students keep the same seat that they had for the first experiment. Do the results of your second experiment support your conclusion?

____

____

____

____

____

____

Name ______________________ Date ____________

YOUR SENSE OF TASTE

Complete these exercises.

A. Can you crack this code?

Below are some body parts that allow you to taste food. Arrange them in the order that shows how they are used as you taste something. Then write the letters onto the blanks above each number. Leave a blank space between words. You will have more blanks than you need.

Body part	Correct order	Code words
1. Brain		1 2 3 4 5 6 7 8 9 10 11 12
2. Taste buds		13 14 15 16 17 18 19 20 21 22 23 24
3. Taste nerves		25 26 27 28 29 30 31 32 33 34 35 36

A secret code word contains these coded letters: 28, 5, 25, 4, 13, 21.

Unscramble them for the secret code word: ______________________

B. The drawing to the right shows an outline of a tongue. Draw in the places that show where each of the four tastes are detected. Label the name of each taste.

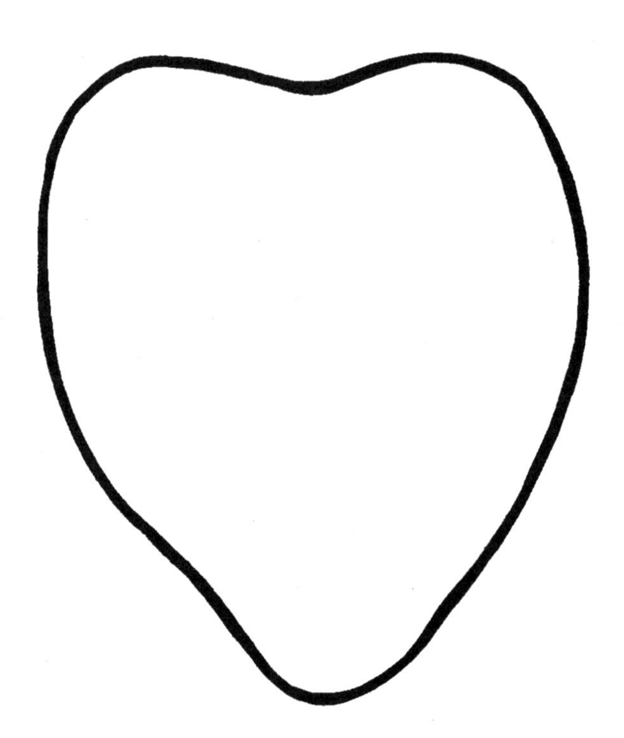

Name ______________________ Date ______________

TEST YOUR SENSE OF TASTE

Materials:
- **blindfold**
- **ten toothpicks**
- **two paper cups of water**
- **two each of five food samples**

(You may wish to use ten different foods and keep the cups covered so that the tasters do not see the foods before they taste them.)

Do This:

1. Your teacher will give you and your partner five food samples, each in a numbered cup.
2. Decide who will be the taster and who will be the recorder. The taster should be blindfolded.
3. Using a toothpick, pick up a small amount of one food sample. Tell the taster to hold his or her nose. Then, place the food on the taster's tongue.

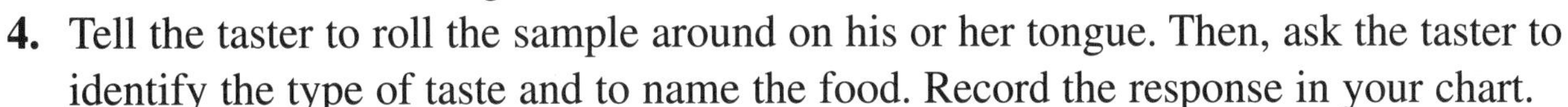

4. Tell the taster to roll the sample around on his or her tongue. Then, ask the taster to identify the type of taste and to name the food. Record the response in your chart.
5. Repeat steps C and D with each food sample. Have the taster drink some water after each sample to clear the taste buds. Use a new toothpick each time.
6. Switch places with the taster, and repeat the activity.

	Food Sample	Food Identified
1.	____________	____________
2.	____________	____________
3.	____________	____________
4.	____________	____________
5.	____________	____________

	Food Sample	Food Identified
1.	____________	____________
2.	____________	____________
3.	____________	____________
4.	____________	____________
5.	____________	____________

Go on to the next page.

Name ______________________ Date ____________

TEST YOUR SENSE OF TASTE, P. 2

Answer these questions.

1. Were there any food samples that the tasters could not identify?

____

____

____

2. Which ones did each taster guess correctly?

____

____

____

3. Why might a taster have difficulty identifying some of the foods?

____

____

____

4. Would the results be the same if the tasters could see the food? Why or why not?

____

____

____

5. Would the results be the same if the tasters did not hold their noses? Why or why not?

____

____

____

Name ______________________ Date ______________

YOUR SENSE OF HEARING

Complete these exercises.

A. Pamela is listening to an orchestra. For her to hear the music, many things happen in her body. Below is a scrambled list of some of these events. Rewrite the events in order from the first thing that happens to the last.

The vibrating eardrum passes the message to nearby bones. Auditory nerves carry the message to the brain. Sound vibrations caused by the instruments reach the eardrum. Bones vibrate and pass the message to the auditory nerve.

____

____

____

____

____

____

____

B. Do you have a favorite sound? Choose your favorite sound. If you don't have one, choose a sound that you dislike.

1. Write a two-line poem that describes this sound. Here is an example:

One of my favorite things
Is to hear the sweet song that my parakeet sings.

____

____

2. If you cup your hands behind your ears when you listen to the sound you described, do you think the sound would be louder or softer? ____________

Explain your answer. ______________________________

____

____

____

Name ______________________ Date ______________

ARE TWO EARS BETTER THAN ONE?

To find out, try this activity.

Materials:
- **blindfold** (A roll of white crepe-paper may be used to make disposable blindfolds.)

Do This: 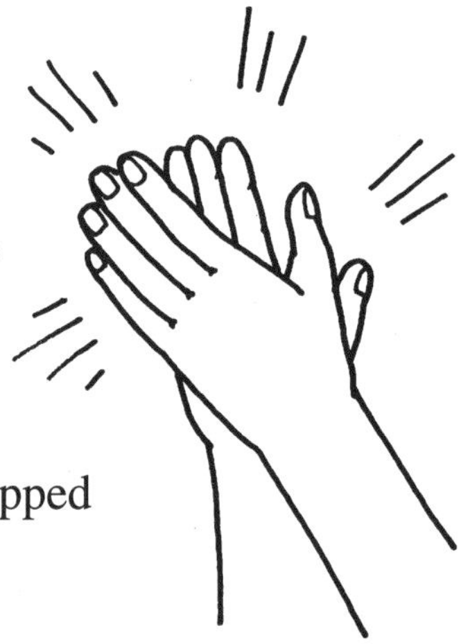
1. Have a partner sit down and put on a blindfold.
2. At a distance of about 3 m from your partner, clap your hands. Ask your partner to point to where the sound came from. Do this ten times from different locations. Record the number of right and wrong responses in your chart.
3. Have your partner press a cupped hand tightly over his or her right ear. Repeat step 2.
4. Now, have your partner uncover the right ear and press a cupped hand over the left ear. Repeat step 2.
5. Switch places with your partner and repeat the activity.

Student 1

	Right	Wrong
Ears uncovered		
Right ear covered		
Left ear covered		

Student 2

	Right	Wrong
Ears uncovered		
Right ear covered		
Left ear covered		

Go on to the next page.

Name ______________________ Date ____________

ARE TWO EARS BETTER THAN ONE?, P. 2

Answer these questions.

1. How many correct responses did you and your partner make with both ears uncovered?

____

____

2. How many did you make with the right ear covered?

____

____

3. How many did you make with the left ear covered?

____

____

4. Is it easier to locate the direction of a sound when listening with both ears or with just one ear? Explain your answer.

____

____

____

____

____

____

____

____

____

Name ______________________________ Date ______________

YOUR SENSE OF SIGHT

This exercise will help you learn about how you see. You may use a science book or an encyclopedia to help answer the questions.

A. Names of parts of your body that help you see are scrambled below. Unscramble each term and rewrite it in the space provided.

1. SLEN ____________________
2. SIIR ____________________
3. POTIC REVEN ____________________
4. TERANI ____________________
5. PLIPU ____________________

Select the term above that best fits each of these definitions.

6. ____________________ The nerve that carries sight messages to the brain.
7. ____________________ The part of the eye that changes light into a pattern.
8. ____________________ The opening in the center of your iris.
9. ____________________ The colored part of the eye.
10. ____________________ The back part of the eye where images are focused.

B. Write two sentences that compare the pairs of terms below:

1. Farsightedness and nearsightedness.

____

____

____

2. The human eye and a camera.

____

____

____

____

Name ________________________________ Date ________________

HOW DOES LIGHT AFFECT THE PUPILS OF YOUR EYES?

To find out, try this activity.

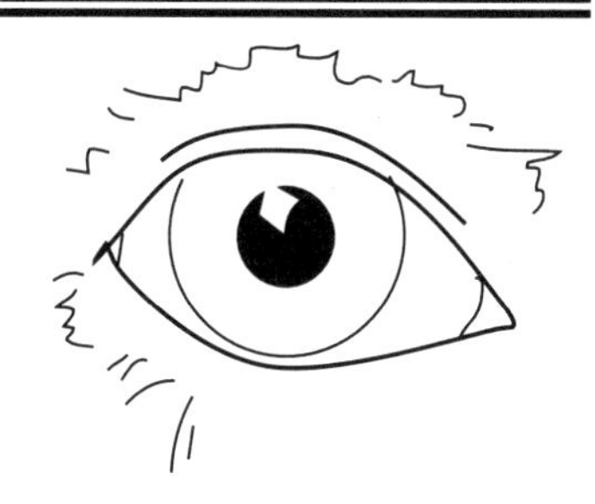

Materials:
- **mirror**

Do This:

1. In normal light, look at your eyes in a mirror. Notice the size of the pupils.
2. Look at the chart of pupil sizes. Find the circle closest in size to your pupil. Circle the letter of the circle. Under it, write *Normal Light.*

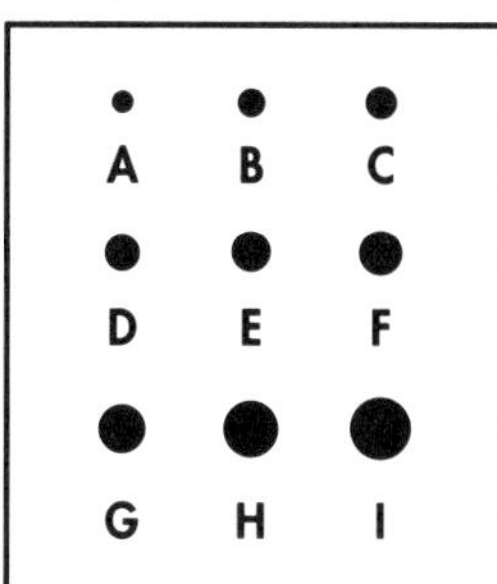

3. Your teacher will dim the lights in the classroom. Wait ten seconds. Then look again at your eyes in the mirror.
4. Circle the letter of the circle closest in size to your pupil now. Under it, write *Dim Light.*
5. Your teacher will turn the lights back on. After ten seconds, look at your eyes again. Again, circle the letter of the circle closest in size to your pupils. Under it, write *Normal Light.*

Answer these questions.

1. What happened to the size of your pupils when the lights were dimmed?

____

2. What happened when the lights were turned back on?

____

3. How do the pupils of your eyes react to light?

____

4. Find out why your pupils do this. What did you find out?

____

____

____

____

Name ______________________ Date ______________

SEEING AN AFTERIMAGE

The lens in your eye changes light into a pattern, or image. The image is carried to your brain by the optic nerve. When you look at an object quickly and then look away or shut your eyes, the image of the object remains for a short time, about $^1/_{16}$ of a second. For example, if you look at a candle and then move the candle away, you will see the candle for a short time after it was moved. Scientists call this an *afterimage*. In this activity, you will see an afterimage.

Materials:

- **index card**
- **black marking pencil**
- **tape**

Do This:

1. Draw a tree with branches on one side of the card.
2. Turn the card over and hold it up to the light. Draw dots on the blank side of the card where you see the tops of the branches.
3. Place the card on your desk and draw a bird at each dot.
4. Tape the card to the top of the pencil.
5. Roll the pencil back and forth quickly between your hands. As you do this, look at the card.

Answer these questions.

1. What did you see? Why? ______________________

____

____

____

2. On what part of the eye does the image form? ______________________

3. A piece of motion picture film is made of many different pictures, which are shown very rapidly. How does a film show motion? Use the word *afterimage* in your answer.

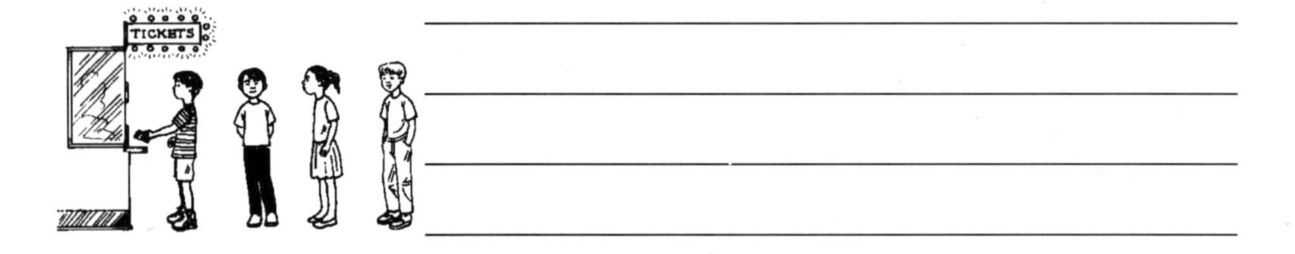

Name ________________________________ Date ______________

Your Sense of Touch

The sense of touch tells us when our body has made contact with another object. The sense of touch can tell us the shape and texture of an object, and it tells us about temperature, pain, and pressure.

Your skin is a sense organ. Within our skin are nerve endings that give us information about touch. There are more nerve endings, or receptors to touch, in some parts of the body than there are in others. Some receptors are deeper in the skin than others. If you touch your skin lightly, you feel touch. If you add pressure, you will stimulate the receptors deeper in the skin. Then you will feel pain.

There are many receptors in the skin. Each receptor senses only one kind of message. The information is carried by the nerve cells to the spinal cord, and then to the brain. The brain sends signals to the body that tell it what to do.

Answer these questions.

1. Your skin has five types of sense receptors. Can you write the names of all of them?

   a. ______________ **c.** ______________ **e.** ______________

   b. ______________ **d.** ______________

2. Look at the drawing. Label the brain, spinal cord, and nerves.

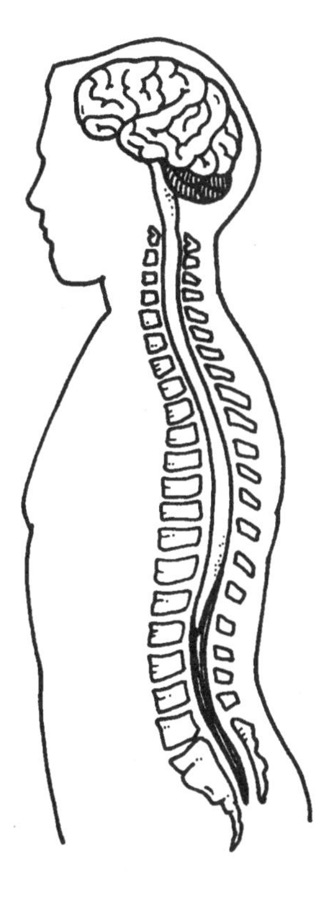

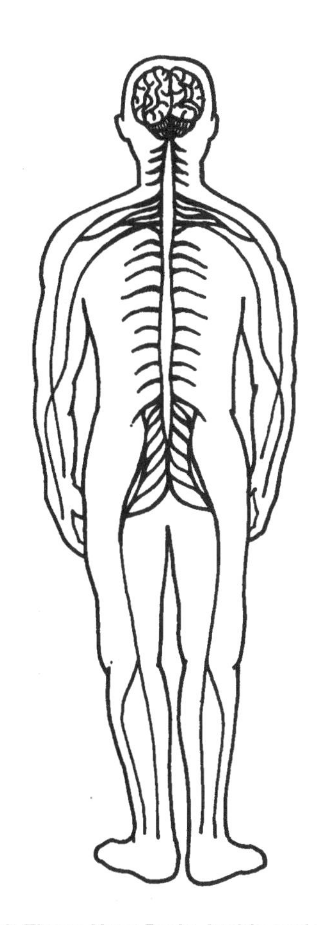

Name ______________________ Date ____________

Distance Between Nerve Cells

Your skin, along with your hair and nails, makes up the integumentary system. This system provides a protective layer for your body.

Your skin contains nerve cells or neurons. These nerve cells give you your sense of touch. Do all areas of your skin have the same number of nerve cells? Can you feel things as well on every part of your body?

Materials:

- **scarf**
- **two toothpicks**
- **ruler**

Do This:

1. With the scarf, blindfold one of your classmates. Hold the two toothpicks together and poke your classmate *gently* on the fingertip until he or she can feel the pressure. It should feel as if there is only one toothpick. Then move the toothpicks slightly apart and try it again. How many toothpicks did your classmate feel? ______________ Keep moving the toothpicks apart until your classmate feels the touch of two toothpicks.
2. Measure the distance between the two toothpicks. This distance is the distance between two nerve cells. Record the number in the chart.
3. Repeat the activity on the back of the hand, the cheek, and the back of the neck. Record your measurements in the chart.

Location	Distance Between Toothpicks
Fingertip	
Back of Hand	
Cheek	
Back of Neck	

Answer these questions.

1. Where are the nerve cells farthest apart? ______________

   Why do you think they are so far apart? ______________

   ____
2. Where are the nerve cells closest together? ______________

   Why do you think they are so close? ______________

   ____

Name ________________________________ Date ______________

How Does Skin Protect Against Infection?

Bright and shiny apples—red, green, and yellow. The "skin" of an apple is more than just something colorful to look at. Find out how skin is helpful.

Materials:

- **two small paper plates**
- **one rotten apple**
- **crayon or marking pen**
- **plastic knife**
- **two fresh apples**

Do This:

1. Label one of the paper plates *uncut skin*. Label the other one *cut skin*.

2. Place a fresh apple on each paper plate.

3. Cut a badly spoiled piece off the rotten apple.

4. Cut a small piece of skin off the apple labeled *cut skin*. Rub the piece of the rotten apple on the area where the skin has been removed. Some of the rotten apple should stick to the fresh apple.

5. Next, rub the same place of the rotten apple on the apple labeled *uncut skin*.

6. Throw away the small pieces of the apple and the rest of the rotten apple. Clean up your area and wash your hands.

7. Put the paper plates with the apples aside. Observe the cut and the uncut apples each day for one week. Record your observations in the table on the next page.

Go on to the next page.

Name ______________________________ Date ______________

How Does Skin Protect Against Infection?, p. 2

Observations		
Day	**Apple with Cut Skin**	**Apple with Uncut Skin**
1		
2		
3		
4		
5		
6		
7		

Answer these questions.

1. Describe what happened to the cut and uncut apples. How can you explain a difference in the two apples?

____

____

____

____

2. How is your skin like the skin of an apple?

____

____

____

____

Name ______________________________ Date ______________

Sense Organs and the Brain

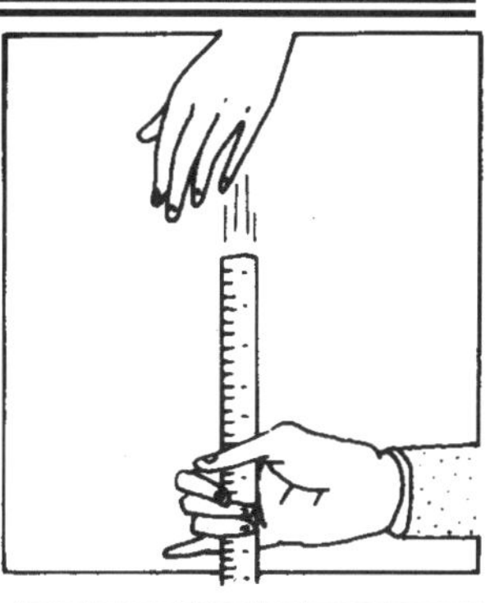

You are learning how your nervous system controls your reactions. In this activity, you and a partner will measure the time it takes for a person to react.

Materials:

- **meter stick**
- **nontransparent tape, such as masking tape**

Do This:

1. Wrap the tape around the stick so that one piece is lined up with the 30 cm mark and the other with the 40 cm mark.
2. One partner should sit with his or her writing arm resting on a desk. The arm should extend beyond the edge of the desk to a point midway between the wrist and elbow.
3. The other partner should stand holding the meter stick at the 100 cm mark. The meter stick should hang within grasp of the seated partner at the 30 cm mark.
4. The seated partner should concentrate on the 40 cm tape mark and be ready to snap his or her fingers shut on that mark when the meter stick begins falling.
5. The standing partner should release the stick without warning. The seated partner should grab the stick. Measure how far the stick fell from the 30 cm mark.
6. Switch places and repeat the above steps.
7. What you measured was the reaction distance. To find the reaction time, use the table shown.
8. Record your results on the chart below.

DISTANCE OF FALL (cm)	TIME OF FALL (s)
1	0.045
2	0.064
3	0.078
4	0.090
5	0.101
6	0.110
7	0.120
8	0.128
9	0.136
10	0.143
11	0.150
12	0.157
13	0.163
14	0.169
15	0.175
16	0.181
17	0.186
18	0.192
19	0.197
20	0.202
21	0.207
22	0.212
23	0.217
24	0.221
25	0.226
26	0.230
27	0.235
28	0.239
29	0.243
30	0.247
31	0.252
32	0.256
33	0.259
34	0.263
35	0.267
36	0.271
37	0.275
38	0.279
39	0.282
40	0.286

	Distance	Time
You		
Partner		

Answer these questions.

1. Who had the faster reaction time? ______________
2. How did your senses and muscles work together in this activity? ______________

Name ______________________ Date ____________

The Skeletal System

There are 206 bones in your skeletal system. The bones form a frame that gives you support. Many bones protect important parts of your body. Some help you move. However, bones can break or become diseased.

Complete these exercises.

A. Go to the library and find books and reference material about the symptoms, causes, and treatments of the following diseases or bone injuries:

bursitis	slipped disk	bone fracture
dislocated shoulder	osteomyelitis	bone break
aplastic anemia	arthritis	bone cancer

B. Write a research report that summarizes what you have learned. Your report should include this information:

1. Causes for the disease or bone injury.
2. Who is likely to get the disease or injury.
3. How doctors treat the disease or repair the injury.

Name ______________________ Date ____________

Bones

This exercise is about the bones in your skeletal system. You may use a science book or an encyclopedia to find the answers.

A. Which two science words belong in the spaces shown in the puzzle? These words name two parts of the skeletal system.

B. Why do people on construction crews or in mines wear helmets? Include the word *cranium* in your explanation.

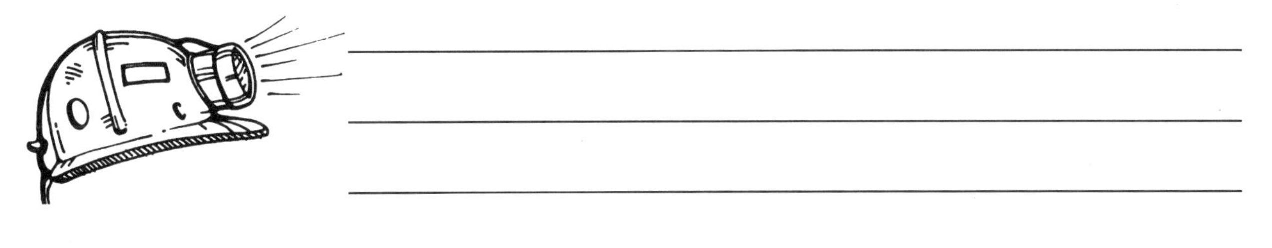

C. Draw a skeleton. Label the cranium, jawbone, backbone, pelvis, and ribs.

Name ______________________ Date ______________

How Bones Join

Each group of drawings and symbols gives you the clues needed to name a type of movable joint. Think of a way to illustrate the third type of movable joint. Then write its name.

1.  ______________________

2. 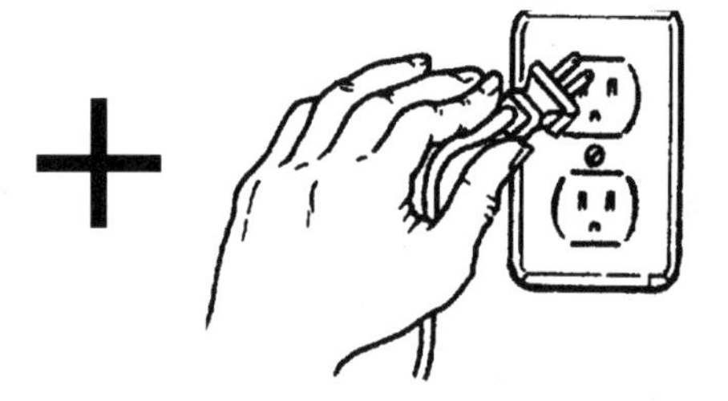 ______________________

3. ______________________

4. Compare the function of the types of joints you listed above.

____

____

____

5. Give an example of a place in your body where each kind of joint is found.

____

____

____

Name ______________________________ Date ______________

THE MUSCULAR SYSTEM

The muscular system produces movement. There are over 600 different muscles in your body. Muscles cover the skeleton. They also line the walls of some organs, such as the heart and stomach. Tendons attach muscles to bone.

Muscles can be voluntary and involuntary. Voluntary muscles are the ones that you can control. You can tell them when to move. Most voluntary muscles are attached to bones. Involuntary muscles, like those of the heart, move without your having to think about them. The muscles that control your eyelids may seem like voluntary muscles. You can blink your eyes when you want to. However, you cannot keep your eyes from blinking when they need to! You do not have complete control over them.

Muscles cause movement by contracting or getting shorter and firmer. This action pulls on the bones or other body structures. Muscles move the blood through your body. They also move food and wastes through your body.

Muscle tone is achieved through exercise. If a person has good muscle tone, the muscles do not completely relax. They are always slightly contracted. For you to have good muscle tone, plenty of blood needs to reach the muscle cells. This requires exercise.

There are three types of muscles in the body. Each type of muscle cell looks different. The **smooth muscles** are long and thin and pointed at each end. The stomach has smooth muscle cells. **Cardiac muscles** make up the heart. They branch out and weave together. **Skeletal muscles** are long and shaped like cylinders (similar to straws). Unlike the other muscle cells, the skeletal muscle cells have many nuclei. The tongue and lips are skeletal muscles, as are the biceps and triceps in your arms.

Match each description with the correct word.

______ **1.** muscles that make up the heart	**a.** tendons
______ **2.** muscles over which you have complete control	**b.** involuntary
______ **3.** what muscles do to cause movement	**c.** contract
______ **4.** necessary for muscle tone	**d.** cardiac
______ **5.** muscle cells with many nuclei	**e.** exercise
______ **6.** attach muscles to bone	**f.** skeletal
______ **7.** muscles that move without conscious effort	**g.** voluntary

Name ______________________________ Date ______________

KINDS OF MUSCLES

Complete these exercises.

A. Each of the following is a description of a type of muscle cell. Rewrite the description under the proper heading below.
Descriptions:
- Muscle cells that branch out and weave together. They make up the heart.
- Long, thin, and pointed cells.
- Long, cylinder-shaped cells.

Skeletal Muscle ______________________________

____

Cardiac Muscle ______________________________

____

Smooth Muscle ______________________________

____

B. Draw a diagram that shows what muscle cells from each of the following body parts would look like. You may use a reference book.

1. Biceps **2.** Heart **3.** Stomach

Name ______________________ Date ____________

How Is Your Muscle Tone?

How strong you are depends on your muscle "tone." Good muscle tone means that your muscle cells are well-nourished. Exercising brings blood carrying food to the muscle cells. In this activity, you will measure the strength of some of your muscles. Work with a partner.

Materials:

- **textbook**
- **watch with a second hand**
- **clear desktop**

Do This:

1. Stretch your left arm out on the desktop so the backs of your upper arm, elbow, lower arm, and hand are all touching the desktop. Ask your partner to put the textbook in your outstretched hand. Grasp the book firmly.
2. Raise the book toward your head. Count how many times you can touch the top of your head with the textbook in 30 seconds. Record your data.
3. Rest for 1 minute. Repeat with your right arm. Then have your partner do the activity.

Number of lifts	**Left arm** ______	**Right arm** ______

Answer these questions.

1. Study your data. Which of your arms had the best muscle tone?

   ____

2. Make a hypothesis that explains any differences between the strength of your right and left arms.

   ____

   ____

3. What exercises could you do to strengthen the muscles of your arms? ______________________________

4. Would it be easier to improve muscle tone for voluntary muscles or involuntary muscles? Explain.

   ____

   ____

Name ______________________________ Date ______________

How Fit Are You?

If you exercise regularly, you can keep your body healthy. Bodies that are totally fit have four important characteristics: they are strong, flexible, well-coordinated, and can endure exercise over a long period of time.

You can do these tests to see how fit you are. Wear loose clothes and gym shoes.

1. Are your arms and shoulders strong?

Flexed-Arm Hang

Using an overhand grip, hang with your chin above the bar and with your elbows flexed. Keep your legs straight and feet free of the floor.

To pass: Hold at least 3 seconds.

Pullups

Using an overhand grip, hang with your arms and legs fully extended, feet free of the floor. Pull your body up until your chin is higher than the bar. Lower your body until your arms are fully extended. Keep pullups smooth and don't kick your legs.

To pass: Do at least 1 pullup.

2. Are your abdominal muscles strong?

Do this with a partner. Lie on your back with knees flexed, feet one foot apart. With fingers laced, grasp your hands behind your head. Have a partner hold your ankles and keep your heels in contact with the floor. Sit up and touch your right elbow to the left knee. Return to the starting position. Then sit up and touch your left elbow to the right knee.

Go on to the next page.

Name ______________________ Date ______________

HOW FIT ARE YOU?, P. 2

To pass: Check the chart below.

Ages	Amount of Exercise
10	25 situps
11	26 situps
12	30 situps

3. Are you well-coordinated?

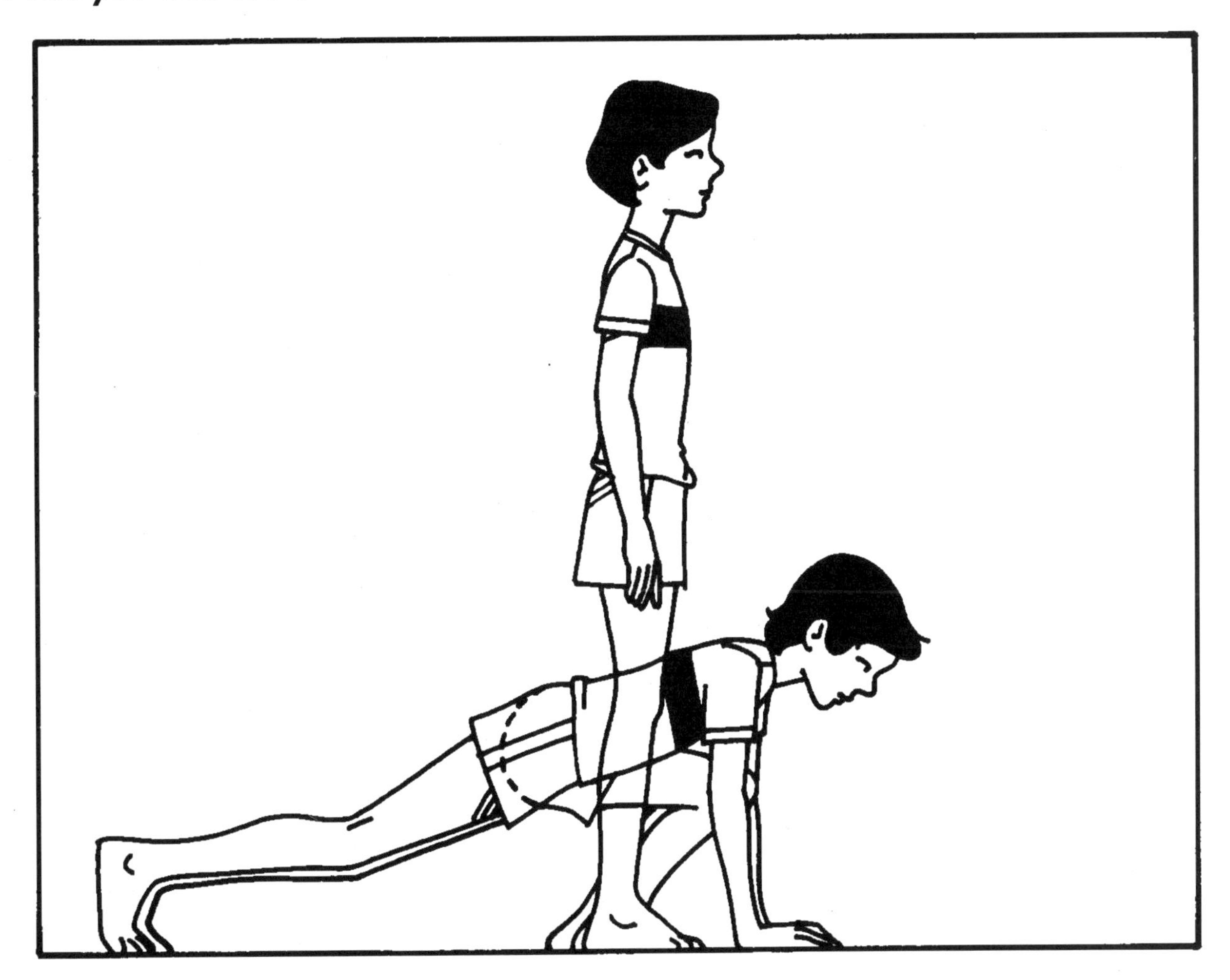

Stand straight. On count 1, bend your knees and place your hands on the floor. On count 2, thrust your legs back so your body is in a pushup position. On count 3, return to a squat position. On count 4, return to a standing position. Do as many as you can in 10 seconds.

To pass: You should do at least 4 squat thrusts in 10 seconds.

Name ______________________________ Date ______________

The Circulatory System

The circulatory system and the respiratory system work together to bring oxygen and nutrients to the body cells and to remove carbon dioxide from the cells.

To follow the blood through the circulatory system, start in the heart—the right atrium, to be exact. As one valve opens, blood that needs oxygen flows from the heart. The valve closes, and another valve opens, allowing the blood to proceed to the lungs. In the lungs, the blood cells get the oxygen they need. They also get rid of carbon dioxide. As red blood cells take in oxygen and give up carbon dioxide, they change in color from dark red to bright red. The blood then leaves the lungs and passes through the heart again—this time through the left ventricle. The heart pumps it through the large arteries into the smaller arteries and capillaries throughout the body. There, oxygen and nutrients are distributed to all the other cells, and wastes are picked up. The blood becomes dark red again. Then the blood returns to the heart—the right atrium—to begin its trip once more.

The blood is carried away from the heart in blood vessels called *arteries*. The blood returns to the heart in blood vessels called *veins*. The smallest blood vessels, no wider than a hair, are called *capillaries*.

Each time the heart "beats," it pushes blood in two directions at once. Some of the blood goes to the lungs, and some of the blood goes to the rest of the body. If you have ever heard a heartbeat, you know that it makes a "puh-pum" type of sound. The "puh" sound is made when the valves of the heart close and push the blood one way, and the "pum" is the sound of different valves pushing the blood the other way. Each beat of the heart is a double pump. The heart pushes your blood through your body about once every minute.

Draw a line from the descriptions to the correct words.

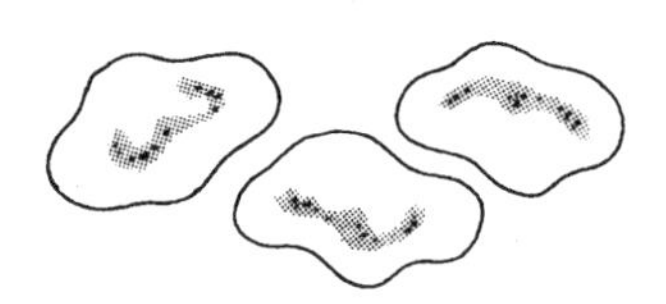

1. brings oxygen and nutrients to cells and takes away carbon dioxide
2. bring blood to the heart
3. take blood away from the heart
4. where cells take in oxygen and give up carbon dioxide
5. the smallest blood vessels
6. works with the circulatory system to deliver oxygen to cells
7. blood with little oxygen
8. blood with plenty of oxygen

a. veins
b. capillaries
c. bright red
d. circulatory system
e. dark red
f. arteries
g. respiratory system
h. lungs

Name ______________________________ Date ________________

FOLLOW THAT BLOOD!

The diagram below shows the circulatory system of the human body. Label the numbered parts of the circulatory system.

Trace the pathway through which blood flows through the system so that it makes a complete loop, beginning and ending at number 1. Use a colored pencil.

1. ______________________
2. ______________________
3. ______________________
4. ______________________
5. ______________________

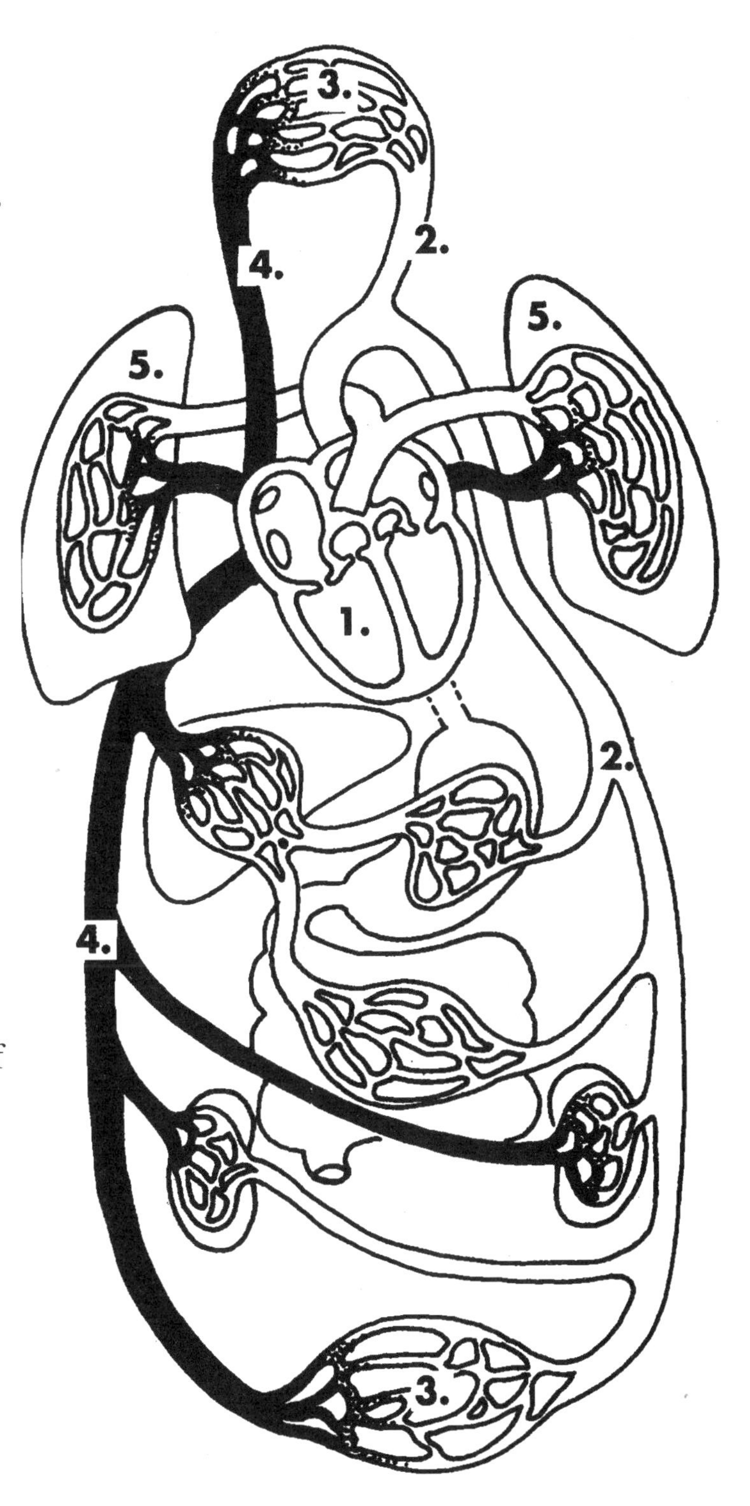

Match each numbered section of the diagram with its description below. Write the number on the line in front of the description.

______ smallest of all blood vessels

______ where red blood cells drop off carbon dioxide and pick up oxygen

______ carries blood away from the heart

______ pumps blood throughout the body

______ carries blood to the heart

Name ______________________________ Date ______________

BLOOD CELLS

Read the following story about blood cells. Then follow the directions.

Red blood cells look like tiny flattened basketballs. Their red color comes from a substance in the cells called *hemoglobin.* Hemoglobin picks up oxygen in the lungs and carries it to all the cells of the body. Sometimes red blood cells move alone in the blood. At other times they travel in rows that look like stacks of coins. Red blood cells are made inside bones. Unlike most cells, a red blood cell has no nucleus. Red blood cells live about four months. Old ones are removed by the white blood cells. One milliliter of blood has between four million and six million red blood cells. If all the red blood cells from an adult's body were placed side by side, they would go around the Earth four times.

White blood cells look different from red blood cells, and they do different work. They surround and destroy invading bacteria. White blood cells are large and contain nuclei. They have irregular shapes. Some are made in the same bones as the red blood cells. Others are made in special glands. Some white blood cells live only a few days. In one milliliter of blood there are between 5,000 and 10,000 white blood cells. When bacteria enter a person's body, the number increases.

Read these statements. Underline the ones that are true.

1. A red blood cell has a nucleus.
2. Red blood cells contain hemoglobin.
3. Red blood cells are made inside bones.

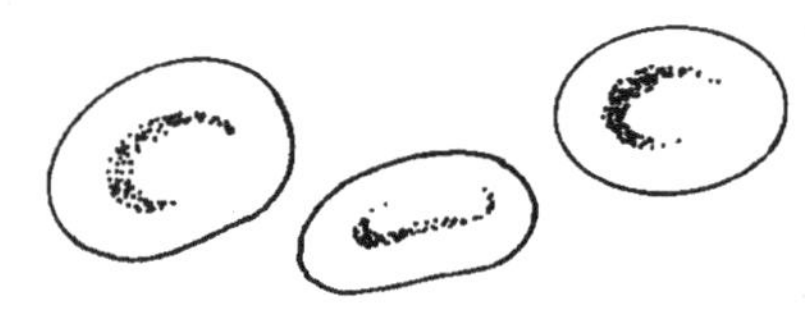

4. You have more white blood cells than red blood cells.
5. White blood cells are larger than red blood cells.
6. Red blood cells remove old white blood cells.
7. White blood cells look like flattened basketballs.
8. Some white blood cells live only a few days.
9. One milliliter of blood may have five million red blood cells.
10. When bacteria enter your body, the number of white blood cells increases.

Name ____________________ Date ____________

CHANGING YOUR PULSE RATE

Your pulse rate does not stay the same all the time. You can find out how it is changed by standing up, lying down, and exercising.

Materials:

- **a clock or a watch with a second hand**

Do This:

1. Stand up for two minutes. While standing, hold your fingers in the correct position for feeling your pulse. Ask a classmate to signal the beginning and end of one minute. Find your pulse rate. Record it in the chart below.
2. Lie down for two minutes. While lying down, take your pulse rate. Record it in the chart.
3. Sit up for two minutes. While sitting up, take your pulse rate. Record it in the chart.
4. Run in place for one minute. As soon as you stop, take your pulse rate. Record it in the chart.

Situation	Pulse Rate
Standing	
Lying Down	
Sitting	
After Running	

Answer these questions.

1. When was your pulse rate the slowest? ____________________
2. When was your pulse rate the fastest? ____________________
3. Why does your pulse rate speed up when you exercise? ____________________

____

____

Name ______________________ Date ______________

The Respiratory System

The respiratory system is responsible for the exchange of gases in the cells of the body. When you inhale, air passes through your nose, down your windpipe, and into two tubes called *bronchial tubes*. These tubes lead into your lungs. The tubes branch many times, like a tree, so that your lungs are filled with tiny tubes. The smallest tubes can only be seen with a microscope. At the ends of these tubes are air sacs.

Air is moved from the air sacs into the cells of the body by diffusion. This is the movement of a substance from an area with a lot of that substance to an area with less of that substance. When the oxygen-poor cells arrive in the lungs from the heart, the oxygen moves into the cells. The carbon dioxide, on the other hand, is more concentrated in the cells, so it moves out of the cells and into the air sacs. When you exhale, the carbon dioxide leaves your body by the same path by which the oxygen entered.

Breathing is only a partly voluntary movement. Part of the reason that you breathe is involuntary. It is caused by the movement of muscles called the *diaphragm*. This is a sheet of muscles beneath your lungs. When the diaphragm moves downward, it increases the space around the lungs, causing air to rush into your lungs. When the diaphragm moves up, it decreases the space around your lungs, and the air rushes out.

Do this crossword puzzle about the respiratory system.

1 2 3 4 5 6 7 8

Across

4. the system that brings oxygen to cells
6. the organ in which the oxygen-carbon dioxide exchange takes place
7. where oxygen goes when it leaves your nose
8. the outside organ that helps you breathe

Down

1. a sheet of muscles below your lungs
2. the tubes that lead into your lung
3. the way oxygen gets into cells
5. microscopic pocket of air in the lungs

Name ______________________________ Date ______________

Respiration Rates

In and out. In and out. Without even having to think about it, you constantly breathe—while you're reading this, while you eat a snack, even while you sleep. With each breath, your body gets the oxygen it needs and gives off carbon dioxide. Find out the number of times you breathe during a day.

Materials:

- **stopwatch, watch, or clock with second hand**
- **calculator**

Do This:

1. Your partner will tell you when to start—and 60 seconds later will tell you to stop.
2. Sit very still. When your partner says "go," start counting your breaths. Remember, breathing in once and then breathing out counts as one breath.
3. Write your number of breaths in the space marked *1 minute* in the table below.
4. Finish filling in the table below. To find out your number of breaths in an hour, multiply the number of breaths in *1 minute* by 60. To find out how often you breathe in a day, multiply the number of breaths in an hour by 24. Multiply that number by 365 to find the breaths in a year.

Number Of Breaths			
1 minute	1 hour	1 day	1 year

Answer these questions.

1. Do you think your breathing rate, or how fast you breathe, can change? Explain your response.

____

____

2. Test your response to the question above. Run in place for 30 seconds. Then repeat steps 1-3. Describe what happens.

____

____

Name ______________________ Date ______________

Are You a Windbag?

You can find out how much air your lungs hold.

Materials:

- **large clear jug**
- **tape**
- **measuring cup**
- **pencil**
- **rubber tubing, about 40 cm long**
- **piece of wax paper**
- **sink or large basin**
- **water**

Do This:

1. Place a strip of tape on the outside of the jug, from top to bottom.
2. Pour 100 mL of water into the jug. Then draw a line on the tape to mark the water level. Repeat this step until the jug is full of water. Label the lines 100 mL, 200 mL, 300 mL, and so on.
3. Fill the sink or basin about half full of water.
4. Fold the wax paper several times. Hold it tightly over the top of the jug. Being careful not to spill any water, turn the jug upside down. Set it on the bottom of the sink.
5. Guide one end of the rubber tubing into the mouth of the jug. Hold a finder over the other end of the tubing.
6. Take a deep breath. Hold it and put the tube in your mouth. Blow through the tube until you cannot blow any more. When you stop to take another breath, hold your finger over the tubing.
7. Observe the water level in the jug. Ask a classmate to record this measurement in the chart on the next page.

Observations

1. Volume of air blown out __________ mL
(water level in milliliters left in the jug)

2. Volume of air always in the lungs __________ mL
(You cannot blow all the air out of your lungs.)

3. Add to find the total volume of air your lungs can hold. __________ mL

How does this activity measure the air in your lungs? As you breathed out, the air pushed the water out of the jug. The amount of water you push out is the same as the amount of air you breathe out.

Name ______________________ Date ______________

THE DIGESTIVE SYSTEM

The digestion of food begins in your mouth. Your saliva contains an enzyme that breaks down starches into sugar. From there, the food moves down the esophagus into the stomach. The stomach continues the digestive process and moves the food to the small intestine. You can show that the digestion of starches begins in your mouth.

Materials:
- **plain, unsalted soda cracker**
- **variety of foods such as bread, unsweetened cereal, peanuts, celery**

Do This:

1. A cracker contains starch. Take a bite of the cracker. How does it taste? ______________________

2. Continue to chew the cracker for one minute.

   How does it taste? ______________________

   Why does it taste this way? ______________________

3. Now test some other foods. How can you find out if they contain starch? ______________________

4. Record your results in the chart below. Compare your results with those of your classmates.

Types of Food	Starch/No Starch
cracker	starch

Name ______________________ Date ______________

The Small Intestine

Stomach muscles push partly-digested food into the small intestine. Most of the food still isn't ready to pass out of the small intestine and into the blood vessels. This food must be broken down further until it can dissolve in water. Only then can food diffuse through the walls of the small intestine. You can make a model of the wall of the small intestine to show that only dissolved substances will pass through.

Materials:

- **paper towel**
- **salt**
- **cinnamon**
- **two clear glasses**
- **funnel**
- **warm water**

Do This:

1. Pour water into one glass. Add two teaspoons of salt. Stir until the salt has dissolved.
2. Add two teaspoons of cinnamon. Stir again.

   Does the cinnamon dissolve? ______________________

3. Fold the paper towel in quarters.
4. Open one side of the paper towel so that it forms a filter. It will act like the wall of the small intestine.
5. Place your filter in the funnel. Place the funnel on top of the empty glass. Pour the mixture through.

   What remains in the filter? ______________________

6. Taste the water that has been filtered.

   How does it taste? ______________________

   What passed through the filter? ______________________

7. What needs to be done in the body to the cinnamon before it can pass through the walls of the small intestine?

   ____

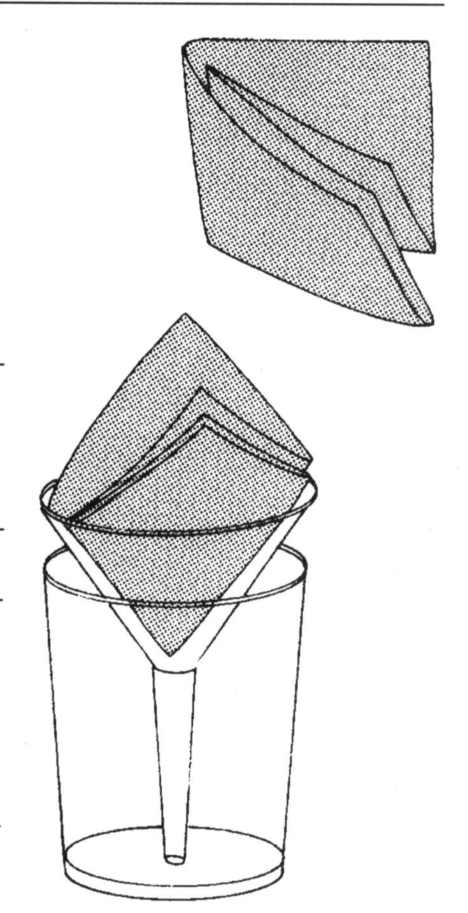

Many foods are like cinnamon. They do not dissolve in water. They must be broken down in the body by digestive juices.

Name ______________________________ Date ______________

DIFFERENT CELLS IN YOUR BODY

Your body has many kinds of cells that do different jobs. Look at these pictures of different kinds of cells.

1\.

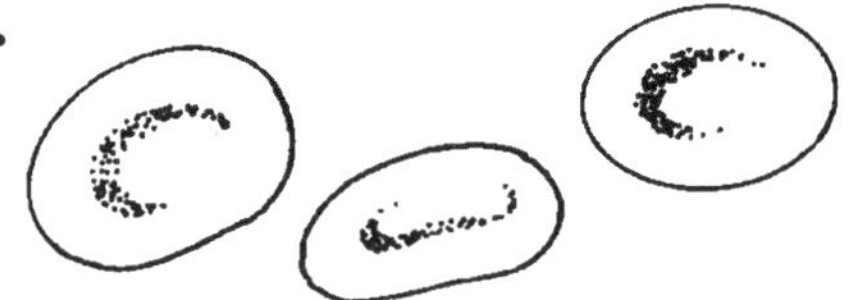

____

2\.

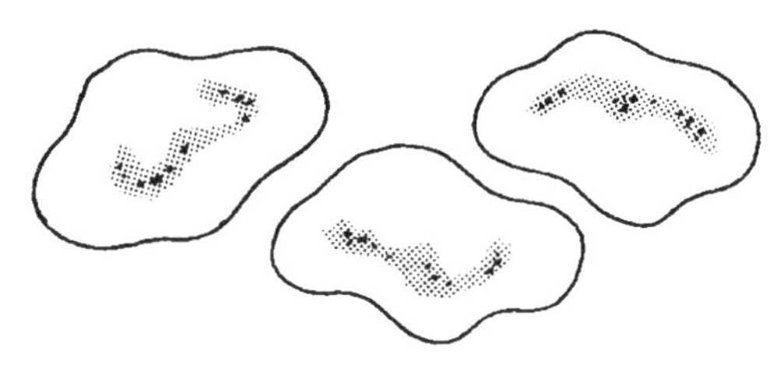

____

3\.

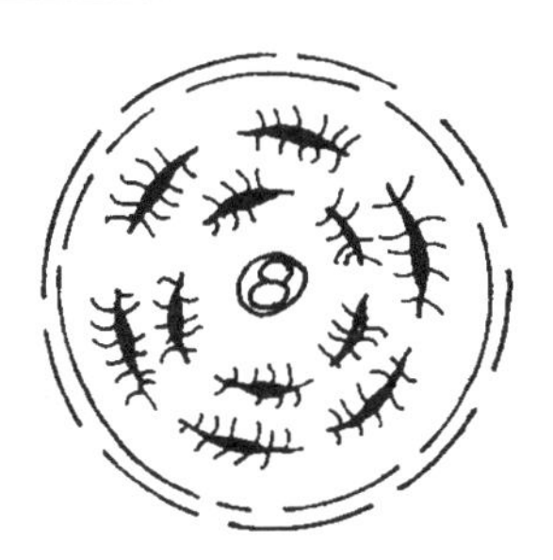

____

4\.

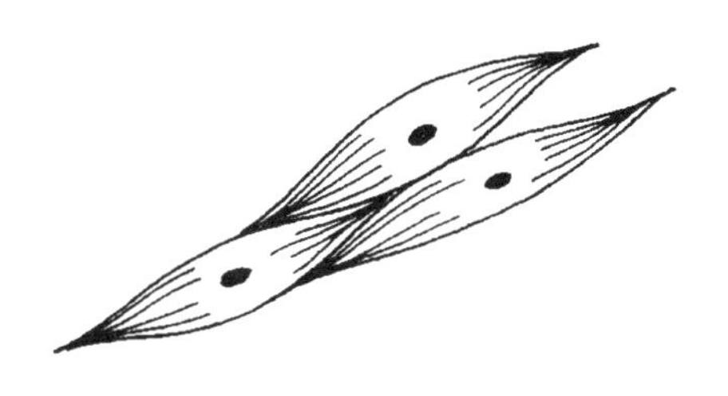

____

5\.

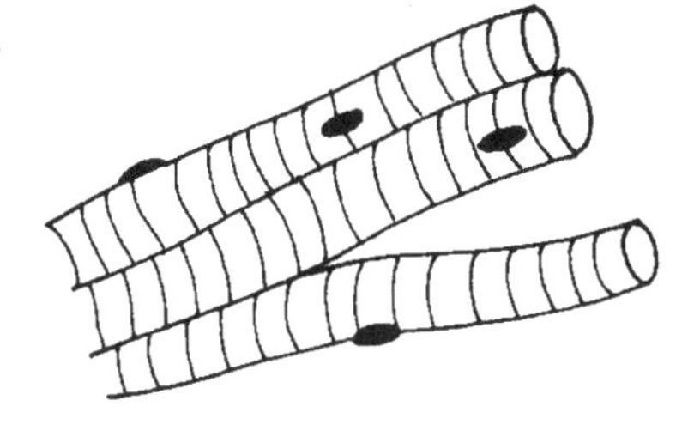

____

6\.

____

7\.

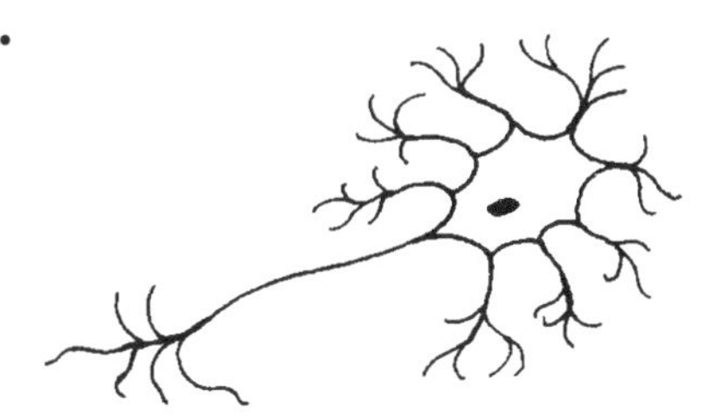

____

8\.

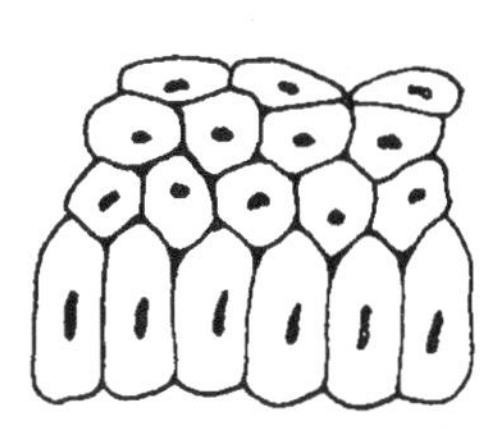

____

Under each picture, write the correct name using the words below. You may use a resource book to look up cells that you do not recognize.

skeletal muscle cells	heart muscle cells	covering cells
white blood cells	red blood cells	neurons
bone cells	smooth muscle cells	

Name ______________________ Date ____________

Reproduction and Heredity

All living things must reproduce, or make more living things like themselves. If a species did not reproduce, all living things of its kind would die out. The reproductive system of humans allows people to make more humans, or to have children.

When people have children, they pass on certain traits and characteristics. This is called *heredity*. Heredity affects the way you look and the way you act. You may have noticed that when adults look at a new baby, they often talk about which parent the baby looks like. This is because the baby has inherited its looks from its parents. As a child grows, there may be times when the child looks more like the mother, and times when the child looks more like the father. There will be certain things the child does that will remind people of the mother or the father, or even of some other relative. All these things are inherited. Other things, such as likes and dislikes and personal fitness, are not inherited. These are the result of the person's lifestyle and environment.

When you study cells, you learn that the nucleus of a cell contains chromosomes. On the chromosomes are genes. Genes determine how offspring will look and act. Each child receives genes from both parents, but some genes are stronger than others are. These genes are called *dominant*. The weaker genes are called *recessive*.

Here is an example. The gene for brown hair is a dominant gene. We say that brown hair is a dominant trait. Blond hair is a recessive trait. If both parents have brown hair, their children will probably have brown hair. If one parent has brown hair, and the other parent has blond hair, the children will still most likely have brown hair, but it is possible for a child to have blond hair. If both parents are blond, then the children will probably be blond.

The combinations can be seen in a chart like this. A brown-haired father may carry a blond-haired gene, because the brown-haired gene will dominate. He may pass on genes like this: Bb

A blond-haired mother cannot have a brown-haired gene. She must pass on genes like this: bb

To see what combinations of genes the children can receive, we can make a chart.

	B	b
b	Bb	bb
b	Bb	bb

In this family, it is possible that half of the children could have blond hair, or there is a 50% chance that a child could have blond hair.

Go on to the next page.

Name ______________________ Date ____________

REPRODUCTION AND HEREDITY, P. 2

If the father had not had a blond gene, he would have passed on genes like this: BB. Now what do the combinations look like? Fill in the chart. How many children can have blond hair now?

	B	B
b		
b		

There are sometimes exceptions to the rules, but they hold true most of the time.

Look at your own family. Can you tell which traits, or characteristics, you received from each of your parents? Do you have the same mannerisms as your father or your mother?

Make a chart about your family. You may wish to consider other close relatives as well. What family resemblances do you see?

Family Member	Hair Color	Eye Color	Height	Shape of Face	Mannerisms You Share

Name ______________________ Date ____________

Sensory Match

On the blank line next to each word or phrase on the left, write the letter of its matching phrase from the column on the right.

1. vibrate ______
2. taste buds ______
3. lens ______
4. cell ______
5. auditory nerve ______
6. pupil ______
7. iris ______
8. spinal cord ______
9. optic nerve ______
10. nervous system ______
11. nucleus ______
12. tissue ______
13. organ ______
14. outer ear ______
15. body system ______

a. the nerve that carries sound messages to the brain

b. groups of organs that work together

c. the sense organs, nerves, brain, and spinal cord

d. a large nerve that carries messages to the brain

e. the opening in the center of the iris

f. group of tissues that works together

g. the part of a cell that controls its activities

h. groups of cells on the tongue

i. the part of the eye that changes light into a pattern

j. the nerve that carries sight messages to the brain

k. tiny living parts of the body

l. to move back and forth

m. group of cells that works together

n. the colored part of the eye

o. the part of the ear that gathers sound vibrations

Name ______________________________ Date ______________

The Food Guide Pyramid

In order to do work, your body needs an energy supply. It gets the energy it needs from the foods you eat. Different kinds of food provide your body with different kinds of energy. Therefore, it is important that your diet be balanced, or made up of various types of foods. The Food Guide Pyramid shows the number of servings of each food group you should eat every day.

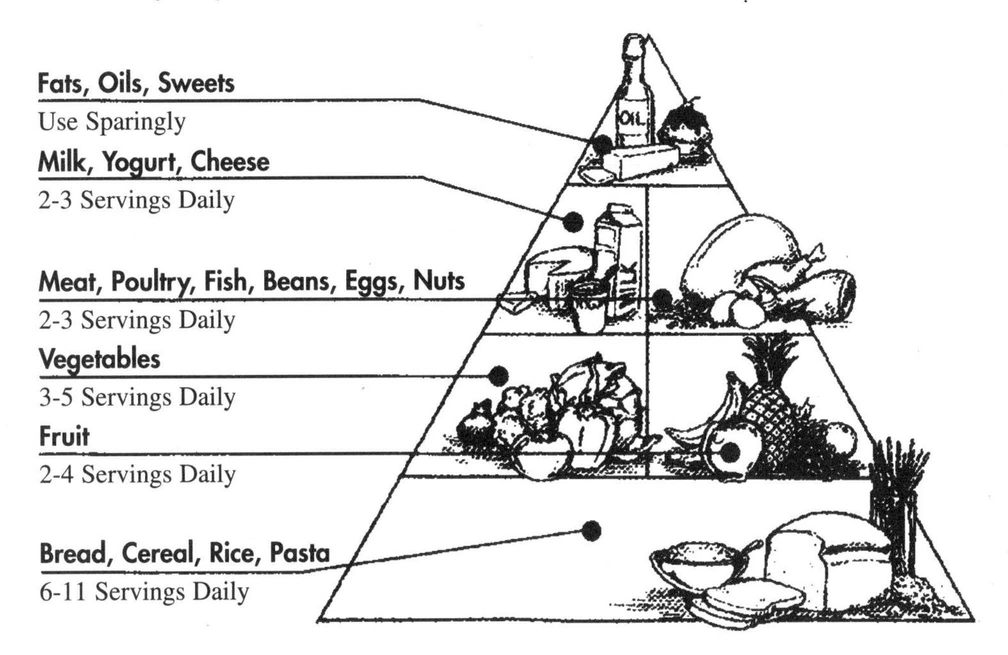

As part of a class project, Laila kept a record of what she ate for one day. Look at the list below and on the next page. Next to each item, write the name of the food group or groups that the item belongs to and the number of servings Laila ate. The first one has been done for you. After you have filled in all the information, answer the questions that follow.

Breakfast	**Food Groups/Servings**
orange juice	fruit, 1
cornflakes	______________
milk	______________
Lunch	
turkey on whole-wheat bread	______________
apple	______________
can of soda	______________

Go on to the next page.

Name ______________________ Date ______________

THE FOOD GUIDE PYRAMID, P. 2

Dinner
spaghetti and meatballs ______________

can of soda ______________

Snacks
ice-cream sandwich ______________

candy bar ______________

Answer these questions.

1. How many more servings of vegetables does Laila need for a well-balanced diet?

____

2. Which other group(s) does she need more servings of in her diet? Explain.

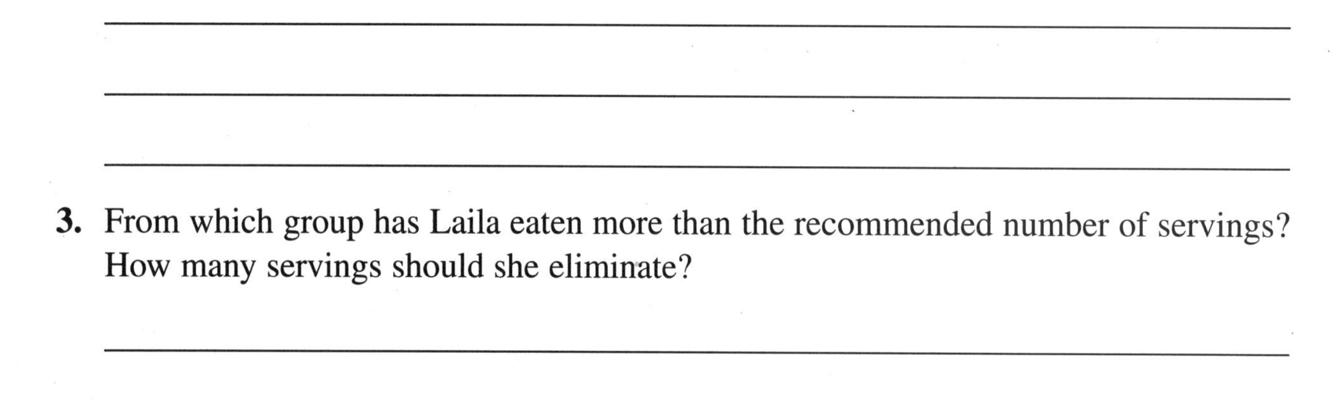

____

____

____

3. From which group has Laila eaten more than the recommended number of servings? How many servings should she eliminate?

____

____

____

4. Keep track of what you eat for one day. Do you have a balanced diet?

Breakfast	Lunch	Dinner	Snacks

Name ______________________________ Date ________________

Good Hygiene

An important part of staying healthy is keeping your teeth and your body clean. It is important to keep the germs and bacteria that can come in contact with your body from spreading. This is why you wash your body and hair. It is very important that you wash your hands whenever you handle garbage or raw meats (such as when you make a hamburger) and each time you use the bathroom. Germs can be spread easily from person to person, too. If you cough or sneeze without covering your mouth, germs fly out into the air and onto other people. If you cover your mouth, the germs are contained, and it is less likely that you will spread your germs to someone else. It is not a good idea to share straws, cups, combs, or hats, either. Germs can be passed from one person to the next this way, as well.

Brushing your teeth keeps bacteria from living in your mouth. They eat small particles of food that are in your mouth. Acids can form that eat into the enamel that protects your teeth. When acid eats into your teeth, decay begins, and you will get cavities. So it is important that you brush your teeth after every meal. Just brushing is not enough, however. If you do not floss, you cannot get all the small food particles from between your teeth, and bacteria will grow anyway. If you get gum disease, the gums will not be able to hold your teeth in place, and they will fall out. Visits to the dentist can help keep your teeth strong and healthy, too. The dentist can see problems with x-rays that can be missed otherwise. Then the dentist can take preventive measures to help your teeth. Brushing, flossing, and visiting the dentist regularly will keep your teeth and gums healthy for many years to come.

Good hygiene helps to keep you healthy and feeling good. It makes you look and smell better, too! Do you have good hygiene habits?

Answer the following questions to see if you need to improve your hygiene.

	Yes	No
1. Do you take a bath or shower every day?	______	______
2. Do you brush your teeth after each meal?	______	______
3. Do you brush your teeth at least twice a day?	______	______
4. Do you floss your teeth every day?	______	______
5. Do you visit the dentist regularly?	______	______
6. Do you cover your mouth when you sneeze or cough?	______	______
7. Do you wash your hands after using the bathroom?	______	______
8. Do you wash your hands after handling garbage or raw meats?	______	______
9. Do you share combs, hats, or other items that go in the hair?	______	______
10. Do you share cups, straws, or other eating utensils?	______	______

If you answered yes to 1-8, and no to 9 and 10, good for you! You have good hygiene habits already! If not, try to improve your habits, and take the test again in two weeks!

Name ______________________________ Date ______________

UNIT 3 SCIENCE FAIR IDEAS

A science fair project can help you understand the world around you better. Choose a topic that interests you. Then use the scientific method to develop your project. Here is an example:

1. **PROBLEM**: Why do children look like their parents?

2. **HYPOTHESIS**: Parents pass on genes to their offspring that determine how the children will look.

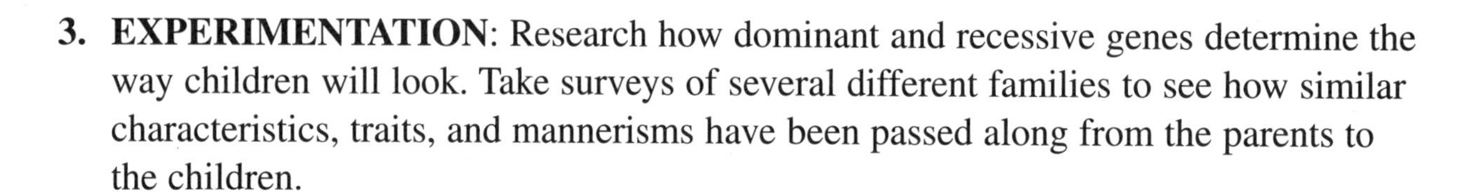

3. **EXPERIMENTATION**: Research how dominant and recessive genes determine the way children will look. Take surveys of several different families to see how similar characteristics, traits, and mannerisms have been passed along from the parents to the children.

4. **OBSERVATIONS**: Many families share common traits and mannerisms. These have been passed from the parents to the children.

5. **CONCLUSION**: Children look like their parents because the parents pass on genes that determine how the children will look.

6. **COMPARISON**: Conclusion agrees with hypothesis.

7. **PRESENTATION**: Display charts showing similarities among the family members that you surveyed. If possible, show photographs of the families and point out similarities.

8. **RESOURCES**: Tell of any reading you did to help you with your experiment. Tell who helped you to get material or set up your experiment.

Other Project Ideas

1. How would eating an unbalanced diet affect a person's health?
2. What happens to muscles that are not used?
3. How does a broken bone heal?
4. What makes the blood stop flowing from a cut?
5. Can a person get too much exercise?

Life Science Grade Five
Answer Key

P. 9 Unit 1 Assessment: 1. c 2. h 3. g 4. d 5. f 6. a 7. e 8. i 9. b 10. l 11. j 12. k

P. 10 Unit 2 Assessment: 1. carbon dioxide 2. amphibians 3. vertebrates 4. Photosynthesis 5. mammals 6. angiosperms 7. metamorphosis 8. bones 9. food chain 10. cold-blooded 11. invertebrates 12. warm-blooded 13. imprints 14. arthropods

P. 11 Unit 3 Assessment: 1. cell 2. tissue 3. organ 4. system 5. systems 6. nervous 7. senses 8. brain 9. bones 10. muscles 11. joints 12. exercise 13. toned 14. blood 15. involuntary 16. voluntary 17. circulatory 18. respiratory 19. lungs 20. carbon dioxide 21. Red 22. White 23. hygiene 24. brushing 25. flossing 26. germs

P. 14 1. Robert Hooke 2. cork 3. a. All plants and animals are made of cells. b. The cell is the basic unit of structure and function in all living things. c. Every cell can reproduce to form new cells.

P. 15 a. cell membrane b. mitochondria c. nucleus d. cytoplasm

P. 16 A. 1= cell wall, 2= cytoplasm, 3=cell membrane, 4= nucleus, 5= chloroplasts, 6= cell membrane, 7= nucleus, 8= cytoplasm B. 1. cell wall, chloroplasts 2. chloroplasts 3. cell wall 4. cell membrane 5. cell a

P. 17 1. Diffusion, Dehydration, Osmosis, nucleus, cell membrane 2. The water that you lose when you sweat comes from the cells of your body. If that water is not replaced by drinking, your cells can become dehydrated. Dehydration can make you very sick, so you should always drink a lot when you sweat a lot. 3. The cells of a wilted plant are dehydrated, just as your body cells are dehydrated when you sweat a lot. Watering a wilted plant is similar to your drinking a lot of water after sweating. The cells of the plant receive water from the soil by the process of osmosis.

P. 18 1. nuclear membrane 2. ribosomes 3. endoplasmic reticulum 4. vacuoles 5. Check students' work for correct order and division process.

P. 19 Check student graphs. Will show cell with nucleus, one organelle, and one mitochondria.

P. 21 Answers will vary.

P. 22 Animal cell only: none; Both plant and animal cells: mitochondria, nucleus, cytoplasm, organelles; Plant cell only: chloroplasts, cell wall

P. 23 1. Some structures are common to both plant and animal cells, and some structures are found in plant cells only. 2. It provides a visual way to see similarities and differences between the two cells. 3. It helps us understand how they work and what their functions are. 4. Since chloroplasts are used in the process of photosynthesis, plant cells can make their own food. Animal cells cannot do this because they lack chloroplasts. 5. Responses will vary but may mention how comparing structure and function helps us gain a better understanding of how organisms work and interact.

P. 24 Students' drawings of plant cell should include cell wall and chloroplasts. Students' drawings of animal cell should not have cell wall or chloroplasts. Carrot Cell: plant, chloroplasts, cell wall; Elephant Cell: animal, cytoplasm, cell membrane, nucleus.

P. 25 Drawings will vary. 4. Answers will vary. Some students may observe that the cell walls are darker. Others may see the nucleus of some of the cells, depending on the microscopes and lighting conditions.

P. 26 Drawings will vary.

P. 27 1. Answers will vary. 2. Answers will vary depending on what protists the students saw. 3. Paramecia use hairlike structures (cilia) to dart about quickly in spiral paths. Euglenas use one long hairlike structure (flagellum) to swim toward light. Amoebas seem to flow as they move by shifting their cytoplasm. 4. Students should conclude that protists exhibit a wide range of colors, sizes, shapes, and structures.

P. 28 1. Amoeba 2. Paramecium 3. Euglena 4. Virus

P. 29 Answers will vary; everything should appear magnified.

P. 31 1. T 2. T 3. F 4. T 5. F 6. T

P. 32 A. The celery should be standing up in the cup, stiff. B. The celery should be limp. C. The celery should be stiff again. 2. limp, no 3. stiff, yes, the water

P. 39 A. 1. Animals breathe in oxygen. 2. Animals breathe out carbon dioxide. 3. Plants take in carbon dioxide. 4. Plants use carbon dioxide, water, sunlight, and their own chlorophyll to make food. 5. Plants give off oxygen. B. Diagrams will vary, but should include the steps in part A.

P. 40 taproot, minerals, stem, food, water, support, leaves, chlorophyll, sunlight, food, photosynthesis

P. 41 1. It has turned the same color as the water. 2. Tubes in the stem have transported the colored water to the flower. 3. Clearly delineated, round strawlike tubes that run vertically through the stem.

P. 42 1. Water formed along the sides. 2. The water from the plant's leaves evaporated because of the heat and condensed on the sides of the jar.

P. 43 1. They were light green or whitish. 2. The leaves of plant **a** were greener. 3. The leaves turned a darker, more healthy green. 4. Chlorophyll; The student knows because chlorophyll causes the green color. 5. The leaves of plant **b** gained chlorophyll by being exposed to the sunlight. This caused the leaves to change color. 6. Plants need sunlight for photosynthesis and the production of chlorophyll.

P. 44 1. a gymnosperm 2. by seeds in cones 3. angiosperms 4. a tiny plant and stored food to help it grow 5. the seed for a fern or moss

P. 45 Top half of page is reproduction by spores. Beginning at the top left and moving clockwise, the current order of the illustrations is third, first, second, fourth. Bottom half of page is reproduction by seeds. Beginning at the top left and moving clockwise, the current order of the illustrations is fourth, first, second, third.

P. 46 1. Angiosperms: tulip, rosebush, reproduce by covered seeds. Gymnosperms: pine tree, spruce tree, reproduce by bare seeds. Ferns: Boston Fern, reproduces by spores. Mosses: haircap moss, reproduces by spores. 2. Mosses do not have real roots and stems. Without roots they cannot dig into the soil for water. A spruce tree needs plenty of water. Its large root system provides the tall tree with support. The large woody stem (trunk) supports it as it grows tall. Mosses have no stem so they do not grow tall.

P. 47 1. fat, rounded, filled out; firm, solid 4. flattened, wrinkled; soft; Roots and stems grew.; inside the bean 5. No roots and stem have grown from this bean.

P. 48 1. herring (or smaller fish) 2. shark 3. the Sun 4. plant plankton 5. Animal plankton would have nothing to eat, so herrings would have nothing to eat, and so on. All the organisms in this food chain would eventually die.

P. 50 1. reproduces 2. 100 3. largest 4. California 5. wet 6. insects 7. fewer 8. blight

P. 51 Invertebrates: starfish, sponge, beetle, spider, lobster, worm, ant, crab, snail, jellyfish, bee, shrimp; Vertebrates: mouse, turtle, fish, frog, lizard, bear, horse, eagle, rabbit, snake, human, dog

P. 52 Posters will vary.

LIFE SCIENCE GRADE FIVE
ANSWER KEY

P. 53 A. 1. A backbone. Vertebrates have backbones. Invertebrates do not. 2. The body would be made of two cell layers with an opening at one end. There would be small holes all over the body. 3. Corals and jellyfish. B. First picture: Flatworm; one body opening, long, flattened body. Second picture: Roundworm; two body openings, pointed at both ends. Third picture: Segmented Worm; body divided into sections, most complicated worm.

P. 54 A. One aquarium should contain: clam, oyster, scallop, mussel, snail, octopus, squid. One aquarium should contain: starfish, sea urchin, sand dollar. B. Drawings will vary, but answer should include the following: The starfish wraps its arms around the mussel. The tube feet pull the shell apart. The starfish pushes its stomach out through its mouth into the open mussel. The mussel's body is digested. The starfish pulls its stomach from the mussel shell.

P. 55 1. A wavelike motion that ripples along the length of the foot. 2. Feelers are located on the snail's head. They alert the snail to the presence of food and potential prey and alert the snail to danger. 3. The snail retreats quickly into its shell. 4. Paragraphs will vary.

P. 56 Answers will vary.

P. 57 4. Answers will vary. 5. Answers will vary. 6. Number will increase.

P. 58 1. insects 2. Female insects can lay thousands of eggs at a time. Even if just a few survive, they will produce many new insects. Insects can feed on almost anything. They have an enormous food supply. 3. Students should label butterfly egg stage, caterpillar, pupa, butterfly shedding its covering, and adult butterfly.

P. 59 1. 3 pairs, or 6 legs 2. It is changing into an adult.

P. 60 Simple Invertebrates: sponges, hollow-bodied invertebrates, worms. Invertebrates with Shells or Spines: mollusks, spiny-skinned invertebrates. Arthropods: mites, ticks, centipedes, millipedes, hard-covered arthropods, spiders, insects

P. 61 1. octopus 2. sponge 3. segmented worm 4. invertebrate 5. squid 6. flatworm 7. vertebrate 8. starfish 9. arthropod 10. insect 11. shrimp or lobster 12. spider, mite, tick 13. moth or butterfly 14. centipede 15. millipede

P. 62 The pictures, clockwise from the top left and ending with the divided circle in the middle should be numbered: 1, 5, 3, 7, 4, 6, 2.

P. 63 1. Tadpoles hatch from eggs. 2. Tadpoles with gills and tails live in water. 3. Tadpoles lose their gills and tails and develop lungs and legs. 4. The animals leave the water and live as frogs. 5. An adult frog lays eggs once again.

P. 64 2. Answers will vary. 3. Answers will vary. 4. The light parts contained food for the newly hatched tadpoles. The dark parts of the egg were the developing tadpoles.

P. 65 1. The fish push themselves through the water with their fins and their tails. They turn by twisting their bodies in the direction in which they are headed. The fish stop by "backpedaling" with their fins. 2. The fish react to the tapping by moving. 3. Again, they react to the tapping by moving.

P. 66 A. 1. cold blooded 2. It laid eggs. 3. thick, scaly skin 4. It had thick skin that kept its body from drying out. B. *Pteranodon*: cold-blooded reptile with scaly skin, hollow bones and wings for flying; laid eggs. Condor: warm-blooded bird with feathers, hollow bones, and wings for flying; lays eggs.

P. 67 1. c 2. e 3. d 4. a 5. b

P. 68 A. 1. The chimpanzee has eyes that face forward. The mouse has eyes that are at the sides of its head. 2. The mouse, because it has chisellike teeth. 3. The chimpanzee, because it has well-developed hands. 4. rodents 5. primates B. Check students' graphs.

P. 69 1, 1, 3, 2, 1, 1

P. 71 1. c 2. d 3. g 4. a 5. h 6. j 7. i 8. b 9. f 10. e

P. 72 Fish: shark; Amphibians: frog; Reptiles: turtle, alligator, snake; Birds: penguin, ostrich, owl; Mammals: cow, squirrel, human, elephant, whale, dog

P. 73 1. cow, because it doesn't lay eggs 2. bird, because it is warm-blooded 3. shark, because it lives in the water 4. dog, because it doesn't have wings 5. worm, because it is an invertebrate 6. shark, because it has a skeleton of cartilage 7. elephant, because it eats plants 8. fish, because it is not a mammal

P. 74 A. 1. trout, frog, snake, kangaroo, human 2. ray, salamander, cat 3. tuna, frog, lizard, human B. 4. primate 5. reptile 6. fish C. Drawings will vary. Students should draw a primate, a reptile, or a fish.

P. 76 1. 10 million years ago 2. camels, three-toed horses, a saber-toothed deer, wading birds, turtles 3. a volcanic eruption 4. It burned their eyes, stuck to their nostrils, tongues, and throats, and eventually suffocated them. 5. They made plaster copies of all the bones.

P. 77 1. B 2. D 3. E 4. C 5. A

P. 79 Drawings will vary.

P. 80 bear, omnivore; sheep, herbivore; eagle, carnivore; elephant, herbivore; owl, carnivore; girl, omnivore (most humans are considered omnivores though some students who are vegetarians may say she is a vegetarian); cow, herbivore; lion, carnivore; giraffe, herbivore; frog, carnivore; raccoon, omnivore; dog, carnivore

P. 82 1. The strip of newspaper in the container with the damp soil decomposed the most. Decomposers in the soil broke down the paper. 2. The strips of newspaper in the dry containers remained dry. Decomposers need moisture in order to function. 3. Decomposers were not present in the sand. 4. They break down materials that would otherwise bury the Earth in waste; they release nutrients back into the food web. 5. Arrows from plants to herbivores and omnivores; Arrows from herbivores to decomposers, scavengers, carnivores, and omnivores; Arrows from omnivores to decomposers, carnivores, and scavengers; Arrows from carnivores to omnivores, decomposers, and scavengers; Arrows from scavengers to decomposers, carnivores, and omnivores; Arrows from decomposers to plants.

P. 86 1. Grassland, 50 cm; Desert, 25 cm; Tropical rain forest, 400 cm; Deciduous forest, 75 cm; Boreal forest, 50 cm; Arctic tundra, 25 cm. 2. tropical rain forest; desert and arctic tundra 3. deciduous forest 4. wet, humid, and very warm 5. plants and animals that require little water, such as cactuses, rattlesnakes, scorpions, and jackrabbits

P. 88 1. arctic tundra, deciduous forests, deserts, boreal forests, grasslands, tropical rain forests 2. top row: arctic tundra, boreal forest, grasslands, bottom row: deciduous forest, tropical rain forest, desert

Pp. 89-90 1. chemical fertilizers, acid rain, landfills, vehicle exhaust 2. a. reuse and recycle materials b. use fewer chemical fertilizers and pesticides c. use materials that can be broken down by organisms in soil and water 3. Responses will vary but should include buildings, roads, crops, animals, plants, and trash. 4. Recycling is important because it helps control pollution. It also saves natural resources, energy, and land.

P. 92 1. Possible responses: People need to get into the habit of recycling; some communities are not set up for recycling; some people may not know or may not care about the benefits of recycling. 2. Possible responses: recycle; not litter; clean up parks or stretches of highway.